Nathaniel L. Britton, Charles Arthur Hollick

The Flora of Richmond County, New York

A Catalogue of the Phlænogamus and Vasculare Cryptogamus Plants...

Nathaniel L. Britton, Charles Arthur Hollick

The Flora of Richmond County, New York
A Catalogue of the Phlænogamus and Vasculare Cryptogamus Plants...

ISBN/EAN: 9783337269968

Printed in Europe, USA, Canada, Australia, Japan

Cover: Foto ©berggeist007 / pixelio.de

More available books at **www.hansebooks.com**

FLORA

OF

RICHMOND COUNTY

NEW YORK.

Flora of Richmond Co., N. Y.—Additions and New Localities, 1880–1882.

Ranunculus aquatilis, L., var. *trichophyllus*, Chaix. Abundant in Clove Lake Swamp; has appeared spontaneously since last year.

Dentaria laciniata, Muhl. Woods near New Dorp. (Miss C. O. Thompson).

Brassica campestris, L., West New Brighton. Introduced in ballast. Also occasional in old fields.

Brassica rapa, L. Occasionally spontaneous in cultivated fields.

Ascyrum Crux-Andreae, L. Kreischerville.

Vaccaria vulgaris, Host. Tompkinsville. (Miss C. O. Thompson.)

Silene inflata, Smith. New Dorp. Rare.

Stellaria longipes, Goldie. Abundant near Port Richmond.

Stellaria uliginosa, Murr. Rossville. Rare.

Sagina decumbens, T. & G. Roadsides near Woodrow.

Sagina decumbens, T. & G., var. *Smithii*, Gray. Garretson's. Rare.

Gypsophila arvensis, L. New Brighton, in gardens and waste-places. Rare. Introduced.

Malva moschata, L. A single plant near Court House. 1880.

Zanthoxylum Americanum, Mill. Port Richmond.

Rhus typhina, L. Richmond Hill.

Medicago sativa, L. West New Brighton. Introduced in ballast.

Desmodium Canadense, DC. Clove Lake.

Tephrosia Virginiana, Pers. Common around Tottenville.

Prunus Mahaleb, Linn. Escaped to roadsides near Garretson's. Rare.

Prunus Cerasus, L. Thoroughly established in woods and copses, probably through the agency of birds.

Pirus communis, L. Sparingly established in woods and borders of fields.

Amygdalus Persica, L. Along old stone walls and open woods.

Amelanchier Canadensis, T. & G., var. *rotundifolia*, Gray. Prince's Bay.

Sedum ternatum, Michx. Roadsides near Woodrow. Rare.

Aethusa Cynapium, L. Near Clove Lake. Rare.

Carum Carvi, L. Roadsides near Four Corners.

Symphoricarpus vulgaris, Michx. Roadsides and waste-places near gardens (escaped).

Lonicera Tartarica, L. Roadside near Huguenot; escaped from gardens.

Viburnum nudum, L., var. *cassinoides*, Gr. Swamps near Watchogue.

Galium trifidum, L., var. *pusillum*, Gr. Near Pleasant Plains. Rare.

Oldenlandia glomerata, Michx. Not rare at Tottenville.

Fedia olitoria, Vahl. Streets of West New Brighton. Rare. (Miss Webster.)

Eupatorium album, L. Common at Tottenville.

Solidago puberula, Nutt. Not rare at Watchogue.

Matricaria Chamomilla, L. West New Brighton. Introduced in ballast.

Artemisia biennis, Willd. Rather plentiful near Clove Lake.

Serratula tinctoria, L. West New Brighton. Probably introduced in ballast.

Senecio aureus, L., var. *obovatus*, Gray. Near Richmond.

Helianthus strumosus, L. Near Clove Lake.

Helianthus decapetalus, L.   Near Clove Lake.
Lysimachia thyrsiflora, L.   Linden Park Swamp.
Mimulus alatus, Ait. · Along a brook near Huguenot.
Galeopsis Ladanum, L.   "Staten Island."   BULL. TORR. BOT. CLUB,
   vi., p. 115.
Perilla ocimoides, L., var. *crispa*, Gray.   "Escaped to roadsides near
   Huguenot."   (W. H. Leggett, in BULL. TORR. BOT. CLUB, ii., 29.)
Echinospermum Lappula, Lehm.   West New Brighton.   Introduced
   in ballast.
Cynoglossum officinale, L.   One plant near Concord.   1880.
Cuscuta compacta, Juss.   Tottenville.
Solanum Carolinense, L.   Waste-places and roadsides, New Dorp
   and New Brighton.
Physalis angulata, L.   New Dorp.   Abundant in one field.
Nicotiana rustica, L.   One plant near Clove Lake.   1880.
Aristolochia Serpentaria, L.   Tottenville.   (Herb. W. H. Leggett.)
Rumex orbiculatus, Gray.   Clove Lake Swamp.
Callitriche Austinii, Engelm.   Tottenville and top of Ocean Terrace.
Croton capitatus, Michx.   Streets of New Brighton.   (Miss C. O.
   Thompson.)
Cannabis sativa, L.   West New Brighton.   Common.   Introduced
   in ballast.   .
Potamogeton pauciflorus, Pursh. "Near Quarantine." (W. H. Leggett).
Sisyrinchium Bermudiana, L., var. *mucronatum*, Gray.   Tottenville.
Smilax tamnoides, L.   "Staten Island, one plant only." (T. F. Allen.
   18th Annual Rept. State Cab. Nat. Hist., p. 202.)
Helonias bullata, L.   In a maple swamp near Kreischerville.
Eleocharis melanocarpa, Torr.   New Springville and New Dorp.
Scirpus planifolius, Muhl.   Tottenville and Garretson's.
Scirpus sylvaticus, L.   Near Garretson's.
Scirpus Eriophorum, Michx., var. *laxus*, Gr.   Near Richmond.
Carex triceps, Michx.   New Dorp.
Carex Muhlenbergii, Schk , var. *enervis*, Boott.   (W H. Leggett, in
   BULL. TORR. BOT. CLUB, vi., 9.)
Carex rosea, Schk., var. *radiata*, Dew.   Near Garretson's.
Carex utriculata, Boott.   Linden Park Swamp.
Carex canescens, L.   Common.
Agrostis spica-venti, L.   West New Brighton.   Introduced in ballast.
Phalaris arundinacea, L., var. *picta*, Gr.   Thoroughly established
   along a brook near Four Corners.
Poa nemoralis, L   West New Brighton.   Introduced in ballast.
Poa serotina, Erhr.   West New Brighton.   Introduced in ballast.
Festuca nutans, Willd.   Hills back of Garretson's.   Not common.
Festuca ovina, L., var. *duriuscula*, Gray.   Near Garretson's.
Bromus multiflorus, Smith.   West New Brighton.   Introduced in
   ballast.
Bromus mollis, L.   New Dorp.   Not common.
Panicum Crus-galli, L., var. *hispidum*, Gr.   Common on the salt
   marshes.
Polypodium vulgare, L.   Sparingly near Silver Lake, Four Corners,
   Richmond, and top of Ocean Terrace.

ARTHUR HOLLICK,
N. L. BRITTON.

OF

RICHMOND COUNTY, NEW YORK.

A CATALOGUE

OF THE

PHÆNOGAMOUS AND VASCULAR CRYPTOGAMOUS PLANTS, WITH
OCCASIONAL NOTES ON THE SAME, GROWING IN
RICHMOND CO., INDEPENDENT OF
CULTIVATION.

BY

ARTHUR HOLLICK,
PORT RICHMOND,

AND

N. L. BRITTON,
NEW DORP.

Price 50 cents. Postage Paid.—Address to either of the above.

STATEN ISLAND
1879.

PREFACE.

In compiling this catalogue our aim has been to give, as completely as possible, a list of all the Phænogamous and Vascular Cryptogamous plants known to grow, or to have grown, in Richmond Co. Many friends have kindly helped us to fill gaps in our list, and to make it as complete as possible up to the present time : but as we shall no doubt have new species come to our notice in the Future, it may become necessary to issue an appendix, and, to render this still more exhaustive, we shall be very grateful for any communication relative to species not mentioned by us.

Wherever the abbreviation T. C. B. occurs it is meant to represent *Torrey Club Bulletin*, the organ of the Torrey Botanical Club.

All species not found personally, we have credited with the name of the persons by whom they were reported to us. The Elliot Collection, referred to often, is a collection made for the late Dr. S. Elliot, of New Brighton, which is very valuable for containing many species whose localities are now destroyed, and the plants themselves extinct.

RANUNCULACEÆ.

Clematis. L. **C. ochroleuca.** Ait. Abundant on Todt Hill.
Also found on Richmond Hill, and near Court House.
C. Virginiana. L. Common.
Anemone. L. **A. Virginiana.** L. Common.
A. nemorosa. L. Common. **Var. quinquefolia,** frequent.
Hepatica. Dill. **H. triloba.** Chaix. Common.
Thalictrum. Tourn. **T. anemonoides.** Michx. Common.
T. dioicum. L. Ocean Terrace, near Garretson's. Woods, near
Egbertville. Not common.
T. purpurascens. L. Rather common.
T. Cornuti. L. Common.
Ranunculus. L. **R. multifidus.** Pursh. Sparingly in ponds
near Court House.
R. alismæfolius. Geyer. Common near Bull's Head, and
New Springville.
R. pusillus. Poir. Wet places ; rather common on the S. side
of the island.
R. abortivus. L. Very common.
R. sceleratus. L. Common.
R. recurvatus. Poir. Common.
R. fascicularis. Muhl. Common.
R. repens. L. Common.
R. bulbosus. L. Rather common.
R. acris. L. Common.
Ficaria. Pers. **F. ranunculoides.** Mön. Escaped into
copses, near Four Corners.
Caltha. L. **C. palustris.** L. Not uncommon.
Coptis. Salisb. **C. trifolia.** Salisb. Near Silver Lake, 1864
(*Elliot Collection*), not since found.--"Ponds south of R. R. between
Clifton and Garretson's." (*I. H. Hall, in T. C. B.*)
Aquilegia. Tourn **A. Canadensis.** L. (*Elliot Collection.*)
Locality destroyed.
A. vulgaris. L. Sparingly escaped from gardens, near Prince's
Bay.
Delphinium. Tourn. **D. Consolida.** L. Escaped along
Clove Road, near the Millpond.
Actæa. L. **A. spicata.** L. **Var. rubra.** Michx. Rare.
Woods, near Silver Lake.
A. alba. Bigel. Common.
Cimicifuga. L. **C. racemosa.** Ell. Common.

MAGNOLIACEÆ.

Magnolia. L. **M. glauca.** L. Swamps, near Giffords and
Watchogue.

Liriodendron. L. **L. Tulipifera.** L. Common.

MENISPERMACEÆ.

Menispermum. L. **M. Canadense.** L. Near Egbertville and Richmond. Not common.

BERBERIDACEÆ.

Berberis. L. **B. vulgaris.** L. Cedar Grove, near New Dorp. Woods, near New Brighton.
Caulophyllum. Michx. **C. thalictroides.** Michx. Woods near Egbertville, in abundance.

NYMPHÆACEÆ.

Brasenia. Schreber. **B. peltata.** Pursh. Silver Lake. abundant
Nymphœa. Tourn. **N. odorata.** Ait. Common.
Nuphar. Smith. **N. advena.** Ait. Common.

SARRACENIACEÆ.

Sarracenia. Tourn. **S. purpurea.** L. Swamp near Gifford's Station. Sparingly.

PAPAVERACEÆ.

Papaver. L. **P. somniferum.** L. Roadside near Four Corners. Sparingly.
Chelidonium. L. **C. majus.** L. Richmond Village. Near New Dorp. Near Egbertville. Not very common.
Sanguinaria. Dill. **S. Canadensis.** L. Abundant in woods near Egbertville, and New Springville.

FUMARIACEÆ.

Dicentra. Bork. **D. Cucullaria.** DC. Near Garretson's. Not common.
D. Canadensis. DC. (*Elliot Collection.*)

CRUCIFERÆ.

Nasturtium. R. Br. **N. officinale.** R. Br. Common in water courses near Clove Lake and Silver Lake.
N. Armoracia. Fries. Escaped along a brook near New Dorp ; also at Tottenville.
Dentaria. L. **D. diphylla.** L. (*Elliot Collection.*)
D. laciniata. Muhl. Bank of Silver Lake. Abundant. Not found elsewhere.
Cardamine. L. **C. rhomboidea.** DC. Not uncommon.
C. hirsuta. L. Common. **Var. sylvatica.** Not uncommon.
Arabis. L. **A. lyrata.** L. Rather common.
A. Canadensis. L. Near Egbertville. Sparingly.

Barbarea. R. Br. **B. vulgaris.** R. Br. Very common.
Sisymbrium. L. **S. officinale.** Scop. Common.
Brassica. Tourn. **B. Sinapistrum.** Bois. Frequent in cultivated fields.
B. nigra. Gray. Common.
Draba. L. **D. verna.** L. Rather common.
Capsella. Vent. **C. Bursa-pastoris.** Mœnch. Too common.
Lepidium. L. **L. Virginicum.** Common.
L. ruderale. L. Common.
L. campestre. L. Becoming very common.
Cakile. Tourn. **C. Americana.** Nutt. Common on the sea beach.
Raphanus. L. **R. Raphanistrum.** L. Very common, especially in oat fields.

CAPPARIDACEÆ.

Polanisia. Raf. **P. graveolens.** Raf. Ravine near Garretson's. Rare.

VIOLACEÆ.

Viola. L. **V. lanceolata.** L. Common.
V. primulæfolia. L. Common, especially near Gifford's and Tottenville.
V. blanda. Willd. Very common.
V. cucullata. Ait. Very common. **Var. palmata.** Gr. Common.
V. sagittata. Ait. Not very common.
V. pedata. L. Very common on Todt Hill, and abundant in other places. An "albino" form was found on Todt Hill; also a red variety near Four Corners.
V. canina. L. **Var. sylvestris.** Regel. Common.
V. striata. Ait. "West New Brighton." (*G. W. Wright.*)
V. delphinifolia. Nutt. "West New Brighton, near Barrett's Dye Works." (*G. W. Wright.*)
V. pubescens. Ait. Common. **Var. scabriuscula.** Torr. and Gr. Occasional.
V. tricolor. L. **Var. arvensis.** Frequent in cultivated ground, near New Dorp.

CISTACEÆ.

Helianthemum. Tourn. **H. Canadense.** Michx. Todt Hill. Richmond Hill. Common.
Hudsonia. L. **H. ericoides.** L. "Tottenville." (*Le Roy, in T. C. B.*)
H. tomentosa. Nutt. Common on the beach.
Lechea. L. **L. major.** Michx. Common.
L. thymifolia. Pursh. Sea beach near New Dorp and Gifford's.
L. Novæ-Cæsareæ. C. F. Austin. Common around Tottenville.
L. minor. Lam. Very common.

8

DROSERACEÆ.

Drosera. L. **D. rotundifolia.** Not common. Near Clove Lake.
D. longifolia. L. (*Elliot Collection.*)

HYPERICACEÆ.

Ascyrum. L. **A. Crux-Andreæ.** L. Near Tottenville (*W. H. L. in T. C. B.*)
Hypericum. L. **H. perforatum.** L. Common.
H. corymbosum. Muhl. Common.
H. mutilum. L. Very common.
H. Canadense. L. Common.
H. Sarothra. Michx. Common.
Elodes. Adans. **E. Virginica.** Nutt. Common.

ELATINACEÆ.

Elatine. L. **E. Americana.** Arnott. "Near Barrett's Dye Works, West Brighton." (*G. W. Wright.*)

CARYOPHYLLACEÆ.

Dianthus. L. **D. Armeria.** L. Common.
D. prolifer. L. On Fort Hill. Scarce.
Saponaria. L. **S. officinalis.** L. Common.
Silene. L. **S. stellata.** Ait. Common.
S. inflata. Smith. "Hillsides on S. I., near the Narrows." (*T. F. A. in T. C. B.*)
S. Pennsylvanica. Michx. Common.
S. Armeria. L. Cultivated ground, near New Dorp.
S. noctiflora. L. On New Dorp Lane.
S. antirrhina. L. "New Brighton." (*G. W. Wright.*)
Lychnis. Tourn. **L. Githago.** Lam. Grain fields. Not very common.
Arenaria. L. **A. serpyllifolia.** L. Very common.
A. squarrosa. Michx. Near Tottenville.
A. lateriflora. L. Quite common.
A. peploides. L. Sea beach, but not common.
Stellaria. L. **S. media.** Smith. Very common.
S. longifolia. Muhl. Tottenville. Rare.
S. borealis. Bigelow. New Springville. Rare.
Cerastium. L. **C. vulgatum.** L. Near Silver Lake. Rare.
C. viscosum. L Very common.
C. nutans. Raf. Woods near Egbertville, and on Ocean Terrace. Abundant.
C. oblongifolium. Torr. Common on nearly all the hills of the Island.
C. arvense. L. New Brighton, 1870. Locality destroyed.
Sagina. L. **S. procumbens.** L. Near Silver Lake, also at Tottenville ; not common.

Spergularia. Pers. **S. rubra.** Presl. **Var. campestris.**
Gray. Not uncommon.
S. salina. Presl. Salt marshes, not so common as the next.
S. media. Presl. Salt marshes. Common.
Spergula. L. **S. arvensis.** L. Cultivated fields, not uncommon.
Anychia. Michx. **A. dichotoma.** Michx. Very common.
Scleranthus. L. **S. annuus.** L. Dry places. Common.
Mollugo. L. **M. verticillata.** L. Common.

PORTULACACEÆ.

Portulaca. Tourn. **P. oleracea.** L. Very common.
P. splendens. L. "Waste places West Brighton and elsewhere." (*W. Chorlton.*)
Claytonia. L. **C. Virginica.** L. Common.

MALVACEÆ.

Malva. L. **M. rotundifolia.** L. Common.
Abutilon. Tourn. **A. Avicennæ.** Gaertn. Common in waste places.
Hibiscus. L. **H. Moscheutos.** L. Common.
H. Trionum. L. Frequent in cultivated ground.

LINACEÆ.

Linum. L. **L. Virginianum.** L. Common.
L. striatum. Walt. Rather common.

GERANIACEÆ.

Geranium. L. **G. maculatum.** L. Common.
G. Carolinianum. L. Common around Tottenville.
G. Robertianum. L. *Elliot Collection.*
Erodium. L'Her. **E. cicutarium.** L'Her. Tottenville, scarce.
Impatiens. L. **I. fulva.** Common.
Oxalis. L. **O. violacea.** L. Common.
O. stricta. L. Very common.

RUTACEÆ.

Ptelea. L. **P. trifoliata.** L. Escaped into copses on Todt Hill, and seemingly well established.
Xanthoxylum. Colden. **X. Americana.** Mill. "New Springville." (*G. W. Wright.*)

SIMARUBACEÆ.

Ailanthus. Desf. **A. glandulosus.** Desf. Common. Spreading rapidly.

ANACARDIACEÆ.

Rhus. L. **R. typhina.** L. Not common. Top of Ocean Terrace.

2

R. glabra. L. Common.
R. copallina. L. Common.
R. venenata. DC. Common in swamps.
R Toxicodendron. L. Very common. A variety of this plant was found in the swamp, at Linden Park, which formed an erect trunk, 6 or 7 feet high, and 2 inches or more in diameter ; having almost entire leaves, thereby seeming to correspond to **R. radicans.** L.

VITACEÆ.

Vitis. Tourn. **V. Labrusca.** L. Common.
V. æstivalis. Michx. Common.
V. cordifolia. Michx. Not very common. Near Court House, and Egbertville.
Ampelopsis. Michx. **A. quinquefolia.** Michx. Common.

RHAMNACEÆ.

Ceanothus. L. **C. Americanus.** L. Common.

CELASTRACEÆ.

Celastrus. L. **C. scandens.** L. Common.
Euonymus. Tourn. **E. atropurpureus.** Jacq. Copses near Tottenville. Common in cultivation.
E. Americanus. L. Woods near Clove Lake. Not common.
Var. obovatus. Torr. & Gray. Frequent.

SAPINDACEÆ.

Staphylea. L. **S. trifolia.** L. (*Elliot Collection.*)
Æsculus. L. **Æ. Hippocastanum.** L. Common in cultivation.
Acer. Tourn. **A. saccharinum.** Wang. Not common. Woods near Egbertville.
A. dasycarpum. Ehrhart. Rather common. Todt Hill, Ocean Terrace, etc.
A. rubrum. L. Common.

POLYGALACEÆ.

Polygala. Tourn. **P. lutea.** L. (*Elliot Collection.*)
P. sanguinea. L. Common.
P. Nuttallii. Torr. & Gray. Near New Brighton and Richmond.
P. verticillata. L. Very common.
P. ambigua. Nutt. Occurs near Tottenville. (If distinct from the latter.)
P. cruciata. L. Not rare. Linden Park Swamp, and along salt meadows near New Dorp.
P. polygama. Walt. (*Elliot Collection.*)

LEGUMINOSÆ.

Lupinus. Tourn. **L. perennis.** L. Mariner's Harbor and Watchogue, abundant. Also at Tottenville.

Crotallaria. I.. **C. sagittalis.** I.. Rare. Near Silver Lake.
Trifolium. I.. **T. arvense.** I.. Common at Tottenville and
Mariner's Harbor.
T. pratense. I.. Common.
T. repens. I.. Very common.
T. agrarium. I.. Common.
T. procumbens. I.. Not common.· Egbertville.
Melilotus. Tourn. **M. officinalis.** Willd. Common around
Tottenville, and sparingly in other places.
M. Alba. Lam. Common. Commences to bloom some time
after the latter.
Medicago. I.. **M. lupulina.** I.. Very common. ·
Amorpha. I.. **A. fruticosa.** I.. **Var. Lewisi.** ·· With
Clematis ochroleuca on S. I." (*I. II. II. in T. C. B.*)
Robinia. I.. **R. Pseudacacia.** I.. Common. Apparently
indigenous.
Tephrosia. Pers. **T. Virginiana.** Pers. Sparingly near
Mariner's Harbor.
Desmodium. DC. **D. nudiflorum.** DC. Common.
D. rotundifolium. DC. Rather common.
D. canescens. DC. Near Court House Station. Not com-
mon. Also near Clove Lake.
D. cuspidatum. Torr. & Gray. Quite common.
D. Dillenii. Darlingt. Common.
D. paniculatum. Very common.
D. ciliare. DC. Near Linden Park Station. Not common.
D. Marilandicum. Booth. Tottenville.
D. viridiflorum. Beck. Tottenville, sparingly.
D. rigidum. DC. Richmond Hill
Lespedeza. M'chx. **L. procumbens.** Rather common.
L. repens. Torr. & Gray. Silver Lake, not common.
L. violacea. Pers. Common, with its varieties.
L. capitata. Michx. Common.
L. hirta. Ell. Near Richmond.
Stylosanthes. Swartz. **S. elatior.** Swartz. Pine Grove,
back of Richmond. Abundant.
Vicia. Tourn. **V. sativa.** I.. Common in cultivated ground.
V. tetrasperma. I.. Not very common. Near Garretson's ;
Richmond Hill.
V. hirsuta. Koch. (*Elliot Collection.*)
V. Cracca. I.. Near West Brighton, abundant.
Lathryus. I.. **L. maritimus.** Bigelow. (*Elliot Collection.*)
L. palustris. I.. **Var. myrtifolius.** Gray. Rather com-
mon along salt meadows, near New Dorp.
Apios. Boerhaave. **A. tuberosa.** Moench. Common.
Phaseolus. L. **P. diversifolius.** Pers. Common along
the shore.
P. helvolus. I.. Quite common.
Amphicarpæa. Ell. **A. monoica.** Nutt. Common. A
perfectly smooth form of this plant was found near Clove Lake.
Galactia. P. Browne. **G. glabella.** Michx. (*Elliot Collection.*)

Baptisia. Vent. **B. tinctoria.** R. Br. Common.
Cassia. L. **C. Marilandica.** L. Quite common near El-
tingville and Giffords. Plenty near Clove Lake.
C. Chamæcrista. L. Common.
C. nictitans. L. Common.
Gleditschia. L. **G. triacanthos.** L. Common in cultiva-
tion, and occasionally escaped.

ROSACEÆ.

Prunus. Tourn. **P. Americana.** Marshall. Near Clove Lake.
Rather rare.
P. maritima. Wang. Common on the beach.
P. Virginiana. L. (*Elliot Collection.*)
P. serotina. Ehrhart. Common.
Spiræa. L. **S. salicifolia.** L. Rather common.
S. tomentosa. L. Quite common. Near Clove Lake, New
Dorp, and Court House.
Poterium. L. **P. Canadense.** Gray. Common. Near Clove
Lake, and along salt meadows near Garretson's and New Dorp.
Agrimonia. Tourn. **A. Eupatoria.** L. Common.
A. parviflora. Ait. Not common. Thickets near New Dorp ;
also at Tottenville.
Geum. L. **G. album.** Gmelin. Common.
G. Virginianum. L. Not common.
Potentilla. L. **P. Norvegica.** L. Common.
P. Canadensis. L. Common. **Var. simplex.** Torr. and
Gr. Common.
P. argentea. L. Not common. Sparingly near Four Corners.
P. Anserina. L. (*Elliot Collection.*)
Fragaria. Tourn. **F. vesca.** L. Common.
F. Virginiana. Ehrhart. (*Elliot Collection.*) Near Silver Lake.
Sparingly.
Rubus. Tourn. **R. triflorus.** Richardson. Common near
Clove Lake.
R. occidentalis. L. Common.
R. villosus. Ait. Very common.
R. Canadensis. L. Common.
R. hispidus. Frequent in swamps.
R. cuneifolius. Pursh. Pine Grove, near Tottenville.
Rosa. Tourn. **R. Carolina.** L. Common.
R. lucida. Ehrhart. Common.
R. rubiginosa. L. Rather common.
R. micrantha. Smith. Rather rare. Huguenot. Near Silver
Lake.
Cratægus. L. **C. Oxyacantha.** L. Abundant on a brook
near New Dorp.
C. Crus-galli. L. Quite common.
C. parvifolia. Ait. Abundant at Tottenville.
Pyrus. L. **P. arbutifolia.** L. Common.
P. malus. L. Is thoroughly established in copses on the Island.

Amelanchier. Medic. **A. Canadensis.** Torr. & Gr.
Var. Botryapium. Gr. Rather common.
Var. oblongifolium. Gr. Common. Typical form not found.

SAXIFRAGACEÆ.

Ribes. L. **R. rotundifolium.** Michx. (*Elliot Collection.*)
R. Grossularia. L. Escaped from cultivation in many places.
R. rubrum. L. Near Princes Bay in woods. Near Egbertville.
Probably escaped.
Parnassia. Tourn. **P. Caroliniana.** Michx. Abundant in
a swamp near Clove Lake.
Saxifraga. L. **S. Virginiensis.** Michx. Common.
S. Pennsylvanica. L. Swamp near Clove Lake. Plentiful.
Heuchera. L. **H. Americana.** L. Common.
Mitella. Tourn. **M. diphylla.** L. (*Elliot Collection.*)
Chrysosplenium. Tourn. **C. Americanum.** Schwein.
Common.

CRASSULACEÆ.

Penthorum. Gronov. **P. sedoides.** L. Common.
Sedum. Tourn. **S. acre.** L. Common in gardens.
S. Telephium. L. Common.

HAMAMELACEÆ.

Hamamelis. L. **H. Virginica.** L. Not rare.
Liquidamber. L. **L. Styraciflua.** L. Common.

HALORAGEÆ.

Myriophyllum. Vaill. **M. ambiguum.** Nutt. Pond near
Court House. Ponds near New Dorp. (*Allen in T. C. B.*)
Proserpinaca. L. **P. palustris.** L. Common.

ONAGRACEÆ.

Circæa. Tourn. **C. Lutetiana.** L. Common.
Epilobium. L. **E. angustifolium.** L. Frequent.
E. palustre. L. **Var. lineare.** Gray. Clove Lake Swamps.
Sparingly.
E. coloratum. Muhl. Common.
Œnothera. L. **Œ. biennis.** L. Common.
Œ. fruticosa. L. Common.
Œ. pumila. L. Near West Brighton. Scarce.
Ludwigia. L. **L. alternifolia.** L. Common.
L. sphærocarpa. Ell. Silver Lake. Abundant in ponds near
New Dorp.
L. palustris. Ell. Common.

MELASTOMACEÆ.

Rhexia. L. **R. Virginica.** L. Common.

LYTHRACEÆ.

Ammania. Houston. **A. humilis.** Michx. Sparingly in a swamp near New Dorp.
Lythrum. L. **L. Hyssopifolia.** L. Abundant on "Red Lane," near New Dorp.
Nesæa. Juss. **N. verticillata.** H. B. K. Common.
Cuphea. Jacq. **C. Viscosissima.** Jacq. Not rare.

CACTACEÆ.

Opuntia. Tourn. **O. vulgaris.** Mill. Common along the South Beach, on Todt hill, and at Tottenville.

CUCURBITACEÆ.

Sicyos. L. **S. angulatus.** S. Common.

UMBELLIFERÆ.

Hvdrocotyle. Tourn. **H. Americana.** L. Rather common.
Sanicula. Tourn. **S. Canadensis.** L. Common.
S. Marilandica. L. Not rare.
Daucus. Tourn. **D. Carota.** L. Very common.
Heracleum. L. **H. lanatum.** Michx. Abundant along a brook near New Dorp.
Pastinaca. Tourn. **P. sativa.** L. Common.
Archemora. DC. **A. rigida.** DC. Abundant in the Linden Park swamp. Also sparingly at Tottenville.
Archangelica. Hoffm. **A. hirsuta.** Torr. & Gr. Common.
A. atropurpurea. Hoffm. Clove Lake Swamp.
Thaspium. Nutt. **T. aureum.** Nutt. Clove Lake Swamp. Abundant.
T. trifoliatum. Gray. Common.
Bupleurum. Tourn. **B. rotundifolium.** L. "On Fort Hill, 1864." (*Dr. F. Hollick.*) Locality destroyed.
Discopleura. DC. **D. capillacea.** DC. Common on the salt meadows. Also in Clove Lake Swamp.
Cicuta. L. **C. maculata.** L. Common.
Sium. L. **S. lineare.** Michx. Common.
Cryptotænia. DC. **C. Canadensis.** DC. Common.
Osmorrhiza. Raf. **O. longistylis.** DC. Near Silver Lake. Along Bard Ave., New Brighton.
O. brevistylis. DC. Common.

ARALIACEÆ.

Aralia. Tourn. **A. racemosa.** L. Near Silver Lake, and Egbertville. Sparingly.
A. hispida. Michx. (*Elliot Collection.*)
A. nudicaulis. L. Common.
A. trifolia. Gray. Frequent.

15

CORNACEÆ.

Cornus. Tourn. **C. florida.** L. Common.
C. sericea. L. Common.
C. stolonifera. Michx. Not rare.
C. paniculata. L'Her. Frequent.
C. alternifolia. L. Richmond Hill. Todt Hill ; rather common.
Nyssa. L. **N. multiflora.** Wang. Common.

CAPRIFOLIACEÆ.

Linnæa. Gronov. **L. borealis.** Gronov. (*Elliot Collection.*)
Lonicera. L. **L. sempervirens.** Ait. Frequent.
Diervilla. Tourn. **D. trifida.** Mœnch. "Todt Hill." (*G. W. Wright.*)
Triosteum. L. **T. perfoliatum.** L. Not rare.
Sambucus. Tourn. **S. Canadensis.** L. Common.
Viburnum. L. **V. Lentago.** L. Common.
V. prunifolium. L. Frequent.
V. nudum. L. Near Giffords. **Var. Claytoni.** Near Watchogue, S. I.
V. dentatum. L. Common.
V. acerifolium. L. Common.

RUBIACEÆ.

Galium. L. **G. Aparine.** L. Rather common.
G. asprellum. Michx. Rather common.
G. trifidum. L. Very common.
G. triflorum. Michx. Common.
G. pilosum. Ait. Common around Tottenville.
G. circæzans. Michx. Common.
Diodia. L. **D. teres.** Walt. Abundant around Tottenville and Clifton.
Cephalanthus. L. **C. occidentalis.** L. Common.
Mitchella. L. **M. repens.** L. Common.
Houstonia. L. **H. cærulea.** L. Very abundant on Todt Hill, and common elsewhere.
Oldenlandia. Plumier. **O. glomerata.** Michx. Sparingly near Richmond.

VALERIANACEÆ.

Valeriana. Tourn. **V. sylvatica.** Richards. A small clump grows in low cedar woods back of Silver Lake.

DIPSACEÆ.

Dipsacus. Tourn. **D. sylvestris.** Mill. Richmond Road near Garretson's in abundance. Roadsides near Bull's Head.

COMPOSITÆ.

Vernonia. Schreb. **V. Noveboracensis.** Willd. Very common. A white variety was found near Richmond in 1876.
Liatris. Schreb. **L. spicata.** Willd. (State Flora and Elliot Collection.)
L. scariosa. Willd. Sparingly, near Clifton.
Eupatorium. Tourn. **E. purpureum.** L. Common.
E. teucrifolium. Willd. Frequent along the salt meadows near New Dorp.
· **E. rotundifolium.** L. Common around Tottenville.
E. pubescens. Muhl. Rossville. (*T. C. B.*)
E. sessifolium. L. Common on Richmond Hill. Woods near New Dorp.
E. perfoliatum. L. Very Common.
E. **ageratoides.** L. Near Silver Lake, along Ocean Terrace. Woods near Richmond.
E. aromaticum. L. Abundant on Richmond Hill, "New Dorp." (*Allen in T. C. B.*)
Mikania. Willd. **M. scandens.** L. Common.
Sericocarpus. Nees. **S. solidagineus.** Nees. Sparingly in woods near Richmond.
S. conyzoides. Nees. Common.
Aster. L. **A. corymbosus.** Ait. Common.
A. macrophyllus. L. Woods near Clove Lake. sparingly. Woods near Eltingville.
A. patens. Ait. Common. **Var. phlogifolius.** Gray. Sparingly on Todt Hill. "S. I." (*Leggett in T. C. B.*)
A. lævis. L. Rather common. **Var. lævigatus.** Sparingly near New Dorp.
A. undulatus. L. Common.
A. cordifolius. L. Common.
A. sagittifolius. Willd. "S. I." (*Leggett in T. C. B.*)
A. ericoides. L. Rather common.
A. multiflorus. Ait. Not common.
A. dumosus. L. Near Tottenville. Not common.
A. Tradescanti. L. Common.
A. miser. L. Common.
A. simplex. Willd. Abundant in Clove Lake Swamp.
A. carneus. Nees. Clove Lake Swamp.
A. puniceus. L. Common.
A. Novæ.—Angliæ. L. Common.
A. acuminatus. Michx. (*Elliot Collection.*)
A. nemoralis. Ait. "S. I." (*Austin in T. C. B.*)
A. flexuosus. Nutt. Rather common in salt meadows.
A. linifolius. L. Common in salt meadows.
A. longifolius. Lam. Not rare.
Erigeron. L. **E. Canadense.** L. Very common.
E. bellidifolium. Muhl. Common.
E. annuum. Pers. Common.
E. strigosom. Muhl. Very common.

17

Diplopappus. Cass. **D. linariifolius.** Hook. Common.
D. umbellatus. Torr. and Gr. Linden Park Swamp. Sparingly, at Tottenville.
Solidago. L. **S. bicolor.** L. Common.
S. latifolia. L. Rather common.
S. cæsia. L. Common.
S. sempervirens. L. Common along salt meadows.
S. neglecta. Torr. and Gray, "S. l." (*T. C. B.*)
S. arguta. Ait. Frequent. **Var. juncea.** Gray. Occasional.
S. altissima. L. Common.
S. ulmifolia. Muhl. Common.
S. odora. Ait. Common.
S. nemoralis. Very common.
S. Canadensis. L. Common.
S. serotina. Frequent.
S. gigantea. Common.
S. lanceolata. L. Very common.
S. tenuifolia. Not very common.
S. patula. Muhl. Near Watchogue.
Chrysopsis. Nutt. **C. Mariana.** Nutt. Common around Tottenville. Near Pleasant Plains. Gifford's.
Inula. L. **I. Helenium.** L. Roadsides. Abundant near Garretson's and New Dorp.
Pleuchea. Cass. **P. camphorata.** DC. Common on salt meadows.
Baccharis. L. **B. Halimifolia.** L. Abundant around a marsh near mouth of New Creek. Near Watchogue. Near Richmond. Tottenville.
Iva. L. **I. frutescens.** L. Common on salt meadows.
Ambrosia. Tourn. **A. trifida.** Common. **Var. integrifolia.** Streets of New Brighton.
A. artemisiæfolia. L. Too common.
Xanthium. Tourn. **X. strumarium.** L. Common. **Var. echinatum.** Gray. Common on the beach.
X. spinosum. L. Waste places, near New Brighton, Stapleton, etc.
Eclipta. L. **E. erecta.** L. "West Brighton." (*G. W. Wright.*)
E. procumbens. Michx. Sparingly near Richmond.
Rudbeckia. L. **R. laciniata.** L. Abundant near Clove Lake Swamp.
R. hirta. L. Common.
Helianthus. L. **H. annuus.** L. Tottenville. Sparingly along the Boulevard, South Beach.
H. giganteus. L. Common.
H. divaricatus. L. Frequent.
H. tuberosus. L. "S. l." (*Leggett in T. C. B.*) Near Gifford's, and New Dorp.
Coreopsis. L. **C. trichosperma.** Michx. Swamps near Watchogue. Also near New Dorp.
Bidens. L. **B. frondosa.** L. Common.

3

B. connata. Muhl. Common in swamps.
B. cernua. L. Clove Lake, abundant.
B. chrysanthemoides. Michx. Common.
B. bipinnata. L. Common.
Helenium. L. **H. autumnale.** L. Common.
Galinsoga. Ruiz and Pav. **G. pavoiflora.** Cav. Becoming a bad weed in streets and gardens in New Brighton. Near Court House Station.
Maruta. Cass. **M. Cotula.** DC. Common.
Anthemis. L. **A. arvensis.** L. Frequent on the South Side.
Achillea. L. **A. Millefolium.** L. Common.
Leucanthemum. Tourn. **L. vulgare.** Lam. Very common.
Tanacetum. L. **T. vulgare.** L. Near Four Corners. Abundant in a field near New Dorp. (Usually near gardens.)
Artemisia. L. **A. caudata.** Michx. Abundant on beach, near Tottenville.
A. biennis. Willd. Stapleton. (*J. W. Congdon in T. C. B.*)
Gnaphalium. L. **G. polycephalum.** Michx. Common.
G. uliginosum. L. Common.
G. purpureum. L. "S. I." (*Leggett in T. C. B.*) Sparingly at Tottenville.
Antennaria. Gærtn. **A. margaritacea.** R. B. Not rare.
A. plantaginifolia. Hook. Common.
Filago. Tourn. **F. Germanica.** L. (*Elliot Collection. State Flora.*)
Erecthites. Raf. **E. hieracifolia.** Raf. Common.
Senecio. L. **S. vulgaris.** L. Weed in gardens near New Dorp.
S. aureus. L. Common. **Var. Balsamitæ.** Common on Todt Hill.
Centaurea. L. **C. Militensis.** L. "Found in a kind of park on S. I." (*J. W. Congdon. in T. C. B.*) (Probably planted.)
C. Cyanus. L. "Waste places." (*W. Chorlton.*)
Cirsium. Tourn. **C. lanceolatum.** Scop. Very common.
C. discolor. Spreng. Common.
C. muticum. Michx. Common in swamps.
C. pumilum. Spreng. Rather common.
C. horridulum. Michx. Common.
C. arvense. Scop. Very common.
Lappa. Tourn. **L. officinalis.** Allioni. **Var. major.** Very common. Frequently white.
Cichorium. Tourn. **C. Intybus.** L. Common.
Krigia. Schreb. **K. Virginica.** Willd. Common.
Cynthia. Don. **C. Virginica.** Don. Quite common.
Leontodon. **L. autumnale.** L. Bard Avenue, New Brighton. (*G. W. Wright.*)
Hieracium. Tourn. **H. scabrum.** Michx. Common.
H. Gronovii. L. Top of Ocean Terrace. Sparingly, near Richmond.

H. venosum. L. Common.
H. paniculatum. L. Common.
H. aurantiacum. Sparingly near Gifford's and near Clove Lake.
Nabalus. Cass. **N. albus.** Hook. Common.
N. Fraseri. DC. Not rare.
N. altissimus. Hook. Near Garretson's. Not common.
Taraxacum. Haller. **T. Dens-leonis.** Desf. Very common.
Lactuca. Tourn. **L. Canadensis.** L. Common. **Var. integrifolia.** Torr. and Gr. Near Richmond.
Mulgedium. Cass. **M. leucophæum.** DC. Rather common.
Souchus. L. **S. oleraceus.** L. Common.
S. arvensis. L. Abundant on the shore near Sailors' Snug Harbor. Also near Vanderbilt's Landing.

LOBELIACEÆ.

Lobelia. L. **L. cardinalis.** L. Common.
L. syphilitica. L. Common. Several white-flowered plants were found in Clove Lake Swamp in 1876 ; these have borne white flowers every year since.
L. inflata. L. Common.
L. spicata. Lam. Rather common. A large patch at Garretson's Station.

CAMPANULACEÆ.

Campanula. Tourn. **C. aparinoides.** Pursh. Swamps, rather common.
C. rapunculoides. L. In a field near New Dorp.
C. glomerata. L. Mill Road, West Brighton (*W. Chorlton.*)
Specularia. Heister. **S. perfoliata.** ADC. Common.

ERICACEÆ.

Gaylussacia. H. B. K. **G. dumosa.** Torr. and Gr. Frequent near Tottenville. **Var. birtella.** "S. I." (*Le Roy, in T. C. B.*)
G. resinosa Torr. and Gr. Common.
Vaccinium. L. **V. stamineum.** L. Frequent.
V. Pennsylvanicum. Lam. Sparingly near Gifford's.
V. vacillans. Solander. Rather common.
V. corymbosum. L. Common, with its varieties.
Chiogenes Salisb. C. **hispidula.** Torr. & Gray. Rare. A small patch near Clove Lake.
Arctostaphylos. Adans. **A. Uva-Ursi.** Spreng. Sparingly near Tottenville.
Epigæa. L. **E. repens.** L. Abundant from Huguenot to Tottenville, near ponds along the beach.
Gaultheria. Kalm. **G. procumbens.** L. Woods near Springville, and Richmond.
Leucothoë. Don. **L. racemosa.** Gray. Frequent.
Andromeda. L. **A. Mariana.** L. Rather common.
A. ligustrina. Muhl. Common.
Clethra. L. **C. alnifolia.** L. Common.

Kalmia. L. **K. latifolia.** L. Common.
K. augustifolia. L. Common at Tottenville and Mariner's Harbor.
Azalea. L. **A. viscosa.** L. Common.
A. nudiflora. L. Very common.
Rhododendron. L. **R. maximum.** L. "Staten Island."
(*Torrey Catalogue, 1819.*) Used to grow at Tottenville.
Pyrola. Tourn. **P. rotundifolia.** L. Quite common.
P. elliptica. Nutt. Common.
Chimaphila. Pursh. **C. umbelleta.** Nutt. Common, but seldom found in flower.
C. maculata. Pursh. Common.
Monotropa. L. **M. uniflora.** L. Common.
M. Hypopitys. L. Rather common.

AQUIFOLIACEÆ.

Ilex. L. **I. opaca.** Ait. Common. Found in flower in cedar woods near New Dorp, and on Richmond Hill.
I. verticillata. Gray. Common.
I. lævigata. Gray. Near Silver Lake and Tottenville.
I. glabra. Gray. Abundant near Clove Lake. Not common elsewhere.

EBENACEÆ.

Diospyros. L. **D. Virginiana.** L. Frequent.

PLANTAGINACEÆ.

Plantago. L. **P. Rugellii.** Michx. Very common.
P. major. L. Frequent.
P. maritima. L. **Var. juncoides.** Gray. Common in salt marshes.
P. lanceolata. L. Very common.
P. Virginica. L. Near Gifford's. Abundant near Egbertville.

PLUMBAGINACEÆ.

Statice. Tourn. **S. Limonium.** L. Salt meadows near Mariner's Harbor. **Var. Caroliniana.** Common on salt meadows.

PRIMULACEÆ.

Trientalis. L. **T. Americana.** Pursh. Rather common.
Lysimachia. Tourn. **L. thyrsiflora.** L. Silver Lake.
L. stricta. Ait. Common.
L. quadrifolia. L. Common. Leaves occur whorled in threes, fours, fives, and sixes, and occasionally opposite and alternate.
L. ciliata. L. Frequent.
L. lanceolata. Walt. Rather common, especially near New Dorp.
L. nummularia. L. Abundant on Todt Hill. Not seen in flower, except on Richmond Road, near Garretson's.

Anagallis. Tourn. **A. arvensis.** L. Common.
Samolus. L. **S. Valerandi.** L. **Var. Americanus.**
Gray. Common.
Hottonia. L. **H. inflata.** Ell. Mariners' Harbor. Abundant in a pond near Court House.

LENTIBULACEÆ.

Utricularia. L. **U. vulgaris.** L. Common.
U. clandestina. Nutt. "Tottenville, S. I." (*Leggett in T. C. B.*)
U. gibba. L. Abundant in a pond near New Dorp, in 1877.

BIGNONIACEÆ.

Tecoma. Juss. **T. radicans.** Juss. Abundant on the beach near Tottenville and Richmond Valley. Roadsides near Egbertville.
Catalpa. Scop. Walt. **C. bignonioides.** Walt. Common in cultivation, and sparingly escaped.
Martynia. L. **M. proboscidea.** Glox. Spontaneous around gardens, New Brighton and Concord.

OROBANCHACEÆ.

Epiphegus. Nutt. **E. Virginiana.** Bart. Common. In one locality near Four Corners the whole plant occurs white.
Aphyllon. Mitchell. **A. uniflorum.** Torr. & Gray. Not uncommon.

SCROPHULARIACEÆ.

Verbascum. L. **V. Thapsus.** L. Common.
V. Blattaria. L. Common.
Linaria. Tourn. **L. Canadensis.** Spreng. Common.
L. vulgaris. Mill. Very common.
Scrophularia. Tourn. **S. nodosa.** L. Not uncommon.
Chelone. Tourn. **C. glabra.** L. Common.
Pentstemon. Mitchell. **P. pubescens.** Solander. Sparingly in a field near West Brighton.
Mimulus. L. **M. ringens.** L. Common.
Gratiola. L. **G. Virginiana.** Common.
Ilysanthes. Raf. **I. gratioloides.** Benth. Common.
Veronica. L. **V. Virginica.** L. Common.
V. Americana. Schweinitz. Rather common.
V. scutellata. L. Common.
V. officinalis. L. Common.
V. serpyllifolia. L. Common.
V. peregrina. L. Common.
V. arvensis. L. Common.
V. Buxbaumii. Tenore. Cultivated ground near New Dorp.
Gerardia. L. **G. purpurea.** L. Common.
G. maritima. Raf. Frequent on salt meadows.
G. tenuifolia. Vahl. Common.
G. flava. L. Not rare.

G. quercifolia. Pursh. Common.
G. pedicularia. L. Not uncommon.
Castilleia. Mutis. **C. coccinea.** Spreng. Very abundant in Clove Lake Swamp. The yellow variety sparingly with it.
Pedicularis. Tourn. **P. Canadensis.** L. Common.
P. lanceolata. Michx. Abundant in the Clove Lake Swamp.
Melampyrum. Tourn. **M. Americanum.** Michx. Very common.

ACANTHACEÆ.

Dianthera. Gronov. **D. Americana.** L. "S. I." (*Le-Roy in T. C. B.*)

VERBENACEÆ.

Verbena. L. **V. hastata.** L. Common. A white flowered variety was found near Gifford's in 1877.
V. urticifolia. L. Very common.
Phryma. L. **P. Leptostachya.** L. Common.

LABIATÆ.

Teucrium. L. **T. Canadense.** L. Rather common.
Trichostema. L. **T. dichotomum.** L. Common.
Mentha. L. **M. rotundifolia.** L. Abundant in Richmond Village. Near Court House station.
M. viridis. L. Common.
M. piperita. L. Common.
M. aquatica. L. **Var. crispa.** Benth. Clove Lake Swamp. Manor Road near Four Corners. Near Court House Station.
M. sativa. L. Garden near New Dorp.
M. Canadensis. L. Common. **Var. glabrata.** Benth. Clove Lake Swamp. Near Tottenville.
Lycopus. L. **L. Virginicus.** L. Rather common.
L. Europæus. L. **Var. sinuatus.** Gray. Common.
Cunila. L. **C. Mariana.** L. Ocean Terrace. Abundant on Todt Hill.
Hyssopus. L. **H. officinalis.** L. (*Elliot Collection*).
Pycnanthemum. Michx. **P. incanum.** Michx. (*Elliot Collection*).
P. clinopodioides. Torr & Gray. Near Clove Lake Swamp.
P. muticum. Pers. Clove Lake Swamp. Low ground near Prince's Bay.
P. lanceolatum. Pursh. Not uncommon.
P. linifolium. Pursh. Common.
Origanum. L. **O. vulgare.** L. (*Elliot Collection*).
Thymus. L. **T. Serpyllum.** L. A small patch on the Turnpike near Silver Lake.
Calamintha. Mœnch. **C. Clinopodium.** Benth. Common.
Melissa. L. **M. officinalis.** L. Clark's Woods, near Court House.

Hedeoma. Pers. **H. pulegioides.** Pers. Very common.
Collinsonia. L. **C. Canadensis.** L. Common.
Monarda. L. **M. punctata.** L. Abundant near Pleasant
Plains.
Lophanthus. Benth. **L. nepetoides.** Benth. (*Elliot Collection*).
L. scrophulariæfolius. Benth. (*Elliot Collection*).
Nepeta. L. **N. Cataria.** L. Common.
N. Glechoma. Benth. Common.
Brunella. Tourn. **B. vulgaris.** L. Common. Found white
occasionally.
Scutellaria. L. **S. pilosa.** Michx. Not uncommon. Abundant near New Dorp and Eltingville.
S. integrifolia. L. Abundant with the last,
S. galericulata. L. Linden Park Swamp.
S. lateriflora. L. Common.
Marrubium. L. **M. vulgare.** L. Around the old house on
the "Point of the Beach," near Gifford's.
Stachys. L. **S. Palustris.** L. **Var. aspera.** Gray. Common. **Var. cordata.** Gray. Sailors' Snug Harbor. **Var. glabra.** Gray. Near Manor Road, West Brighton.
S. hyssopifolia. Michx. Sparingly in Swamps near New Dorp.
Leonurus. L. **L. Cardiaca.** L. Common.
Lamium. L. **L. amplexicaule.** L. Common.

BORRAGINACEÆ.

Lithospermum. Tourn. L. **L. arvense.** L. Abundant
around Tottenville. Near Richmond.
Myosotis. L. **M. palustris.** Withering. **Var. laxa.** Gray.
Common.
M. arvensis. Hoffm. Near Four Corners ; scarce.
M. verna. Nutt. Common.
Cynoglossum. Tourn. **C. Morisoni.** DC. Woods near
Egbertville.
Heliotropium. Tourn. **H. Europæum.** L. Sparingly on
roadsides, New Brighton.

HYDROPHYLLACEÆ.

Hydrophyllum. L. **H. Virginicum.** L. Abundant in
woods near Egbertville.
H. Canadense. L. Sparingly with the last, 1876.

POLEMONIACEÆ.

Phlox. L. **P. paniculata.** L. Escaped from gardens.

CONVOLVULACEÆ.

Ipomœa. L. **I. purpurea.** Lam. Common in waste places.
I. pandurata. Meyer. Sparingly at Tottenville.

Convolvulus. L. **C. arvensis.** L. Abundant in a field near New Dorp.
Calystegia. R. Br. **C. sepium.** R. Br. Common.
Cuscuta. Tourn. **C. inflexa.** Engelm. "S. I." (*W. H. L.* in *T. C. B.*)
C. Gronovii. Willd. Common.

SOLANACEÆ.

Solanum. Tourn. **S. Dulcamara.** L. Common.
S. nigrum. L. Common.
Physalis. L. **P. viscosa.** L. Common.
P. pubescens. L. Near Court House.
P. Alkekengi. L. Cultivated ground near New Dorp. Tottenville.
Nicandra. Adans. **N. physaloides.** Gærtn. (*Elliot Collection*).
Lycium. L. **L. vulgare.** Dunal. Common near dwellings.
Datura. L. **D. Stramonium.** L. Common.
D. Tatula. L. Common.
Nicotiana. L. **N. rustica.** L. Spontaneous in a garden in Concord.

GENTIANACEÆ.

Sabbatia. Adans. **S. angularis.** Pursh. Not rare on Ocean Terrace. Near Clove Lake.
S. stellaris. Pursh. Common on salt meadows.
S. gracilis. Salisb. Salt marsh near New Dorp.
Gentiana. L. **G. crinita.** Frœl. Frequent.
G. Andrewsii. Griseb. Clove Lake Swamp. Near Richmond.
G. Saponaria. L. Clove Lake Swamp. Locality destroyed.
Bartonia. Muhl. **B. tenella.** Muhl. Common.
Menyanthes. Tourn. **M. trifoliata.** L. (*Elliot Collection*).

APOCYNACEÆ.

Apocynum. Tourn. **A. androsæmifolium.** L. Near New Dorp Station. Not rare.
A. cannabinum. L. Common.
Vinca. L. **V. minor.** L. Escaped from old gardens.

ASCLEPIADACEÆ.

Asclepias. L. **A. Cornuti.** Decaisne. Common.
A. phytolaccoides. Pursh. Not rare.
A. purpurascens. L. Rather common.
A. quadrifolia. Jacq. Frequent.
A. incarnata. L. **Var. pulchra.** Common. A white variety was found in Clove Lake Swamp.
A. obtusifolia. Michx. Abundant in pine woods near Tottenville.

A. tuberosa. L. Common.
A. verticillata. L. Abundant around Todt Hill.
Acerates. Ell. **A. viridiflora.** Ell. Rather common.

OLEACEÆ.

Ligustrum. Tourn. **L. vulgare.** L. Frequent.
Fraxinus. Tourn. **F. Americana.** L. Common.
F. pubescens. Lam. Frequent.

ARISTOLOCHIACEÆ.

Asarum. Tourn. **A. Canadense.** L. Frequent.
Aristolochia. Tourn. **A. Serpentaria.** L. (*Elliot Collection.*)

PHYTOLACCACEÆ.

Phytolacca. Tourn. **P. decandra.** L. Common.

CHENOPODIACEÆ.

Chenopodium. L. **C. album.** L. Very common.
C. glaucum. L. Foot of New Dorp Lane. Near Richmond.
C. urbicum. L. S. I. (*T. C. B.*) Richmond Village.
C. murale. L. "Vanderbilt's Landing, S. I." (*W. H. L. in T. C. B.*)
C. hybridum. L. Common.
C. Botrys. L. Frequent.
C. ambrosioides. L. Very common in the streets of New Brighton, &c.
Blitum. Tourn. **B. maritimum.** Nutt. Rare. Salt meadows near New Dorp in 1877.
Atriplex. Tourn. **A. patula.** L. Common.
A. arenaria. Nutt. Common on the beach.
Salicornia. Tourn. **S. herbacea.** L. Common on salt marshes.
S. Virginica. L. Salt meadows near New Dorp.
S. fruticosa. L. **Var. ambigua.** Gray. Sea beach.
Suæda. Forskal. **S. maritima.** Dumortier. Salt marshes. Common near Tottenville.
Salsola. L. **S. Kali.** L. Common.

AMARANTACEÆ.

Amaranthus. Tourn. **A. paniculatus.** L. Waste grounds, New Brighton.
A. retroflexus. L. Common.
A. albus. L. Common.
A. Blitum. L. "Quarantine Landing, S. I." (*W. H. L. in T. C. B.*)
Acnida. L. **A. cannabina.** L. Common.

4

POLYGONACEÆ.

Polygonum. L. **P. orientale.** L. Frequent.
P. Pennsylvanicum. L. Common.
P. incarnatum. Ell. Common.
P. Persicaria. L. Common.
P. Hydropiper. Very common.
A. acre. H. B. K. Common.
P. hydropiperoides. Michx. Common.
P. amphibium. L. Swamp near New Dorp.
P. Virginianum. L. Common.
P. articulatum. L. Common on the beach.
P. aviculare. L. Very common. **Var. erectum.** Roth. Common.
P. ramosissimum. Michx. Sparingly in salt marsh near New Dorp.
P. maritimum. L. Sea Beach. Scarce.
P. tenue. Michx. Not rare.
P. arifolium. L. Common.
P. sagitattum. L. Very common.
P. Convolvulus. L. Oat-field near New Dorp, 1876.
P. dumetorum. L. **Var. scandens.** Gray. Common.
Fagopyrum. Tourn. **F. esculentum.** Mœnch. Common.
Rumex. L. **R. Brittanica.** L. Frequent.
R. verticillatus. L. Rather common.
R. crispus. L. Common.
R. obtusifolius. L. Rather common.
R. Acetosella. L. Very common.

LAURACEÆ.

Sassafras. Nees. **S. officinale.** Nees. Common.
Lindera. Thunberg. **L. Benzoin.** Meisner. Common.

SANTALACEÆ.

Comandra. Nutt. **C. umbellata.** Nutt. Common.

LORANTHACEÆ.

Phoradendron. Nutt. **P. flavescens.** Nutt. Said to have grown in a swamp near Clifton, but not recently found.

SAURURACEÆ.

Saururus. L. **S. cernuus.** L. Common.

CERATOPHYLLACEÆ.

Ceratophyllum. L. **C. demersum.** L. **Var. echinatum.** Clove Lake ; abundant.

CALLITRICHACEÆ.

Callitriche. L. **C. Austini.** Englm. "S. I." (*IV. II. L. in T. C. B.*)
C. verna. L. Common.

EUPHORBIACEÆ.

Euphorbia. L. **E. polygonifolia.** L. Common on the beach.
E. Maculata. L. Very common.
E. Hypericifolia. L. Common.
E. corollata. L. "S. I." (*Le Roy in T. C. B.*)
E. Ipecacuanhæ. L. Common at Mariner's Harbor and Tottenville.
E. Cyparissias. L. Frequent along roadsides.
E. Lathyris. L. Roadside near New Dorp.
Acalypha. L. **A. Virginica.** L. Very common. **Var. gracilens.** Gray. Rather common.

URTICACEÆ.

Ulmus. L. **U. fulva.** Mich. West New Brighton.
U. Americana. L. Common.
Celtis. Tourn. **C. occidentalis.** L. Common.
Morus. Tourn. **M. rubra.** L. Occasional.
M. alba. L. Near Richmond. Frequently planted and occasionally escaped.
Urtica. Tourn. **U. gracilis.** Ait. Frequent.
U. Dioica. L. Richmond Village.
Láportea. Gaudichaud. **L. Canadensis.** Gaudichaud. Woods near Egbertville ; abundant.
Pilea. Lindl. **P. pumila.** Gray. Very common.
Bœhmeria. Jacq. **B. cylindrica.** Willd. Common in low grounds.
Cannabis. Tourn. **C. sativa.** L. Waste places. (*IV. Chorlton.*)
Humulus. L. **H. Lupulus.** L. Frequent along fences.

PLATANACEÆ.

Platanus. L. **P. occidentalis.** L. Common. Not many young trees observed.

JUGLANDACEÆ.

Juglans. L. **J. cinerea.** L. Near Stapleton and West Brighton.
J. nigra. L. Common.
Carya. Nutt. **C. alba.** Nutt. Common.
C. tomentosa. Nutt. Common.

C. porcina. Nutt. Common.
C. amara. Nutt. Ocean Terrace, near Court House Station.
C. microcarpa. Nutt. Near Court House.

CUPULIFERÆ.

Quercus. L. **Q. alba.** L. Common.
Q. obtusiloba. Michx. Sea Beach, near Clifton.
Q. bicolor. Willd. Rather common.
Q. Prinus. L. Quite common.
Q. prinoides. Willd. Tottenville.
Q. nigra. L. Tottenville.
Q. coccinia. Wang. Common. **Var. tinctoria.** Gray.
Rather common.
Q. rubra. L. Common.
Q. palustris. Du Roi. Common.
Castanea. Tourn. **C. vesca.** L. **Var. Americana.**
Michx. Common.
Fagus. Tourn. **F. ferruginea.** Ait. Common.
Corylus. Tourn. **C. Americana.** Walt. Common.
C. rostrata. Ait. "Near Barrett's Dye Works, West New
Brighton." (*G. W. Wright.*)
Carpinus. L. **C. Americana.** Michx. Common.

MYRICACEÆ.

Myrica. L. **M. cerifera.** L. Common.
Comptonia. Solander. **C. asplenifolia.** Ait. Common.

BETULACEÆ.

Betula. Tourn. **B. lenta.** L. Common.
B. alba. Var. populifolia. Spach. Common.
Alnus. Tourn. **A. serrulata.** Ait. Common.

SALICACEÆ.

Salix. Tourn. **S. tristus.** Ait. "S. I." (*T. C. B.*)
S. humilis. Marshall. Common.
S. discolor. Muhl. Very common.
S. sericea. Marshall. Rather common.
S. lucida. Muhl. Frequent.
S. nigra. Marsh. Common.
S. alba. L. **Var. vitellina.** Gray. Common.
S. Babylonica. Tourn. Commonly planted.
S. fragilis. L. "Near Barrett's Dye Works,West New Brighton."
(*G. W. Wright.*)
Populus. Tourn. **P. tremuloides.** Michx. Common.
P. grandidentata. Michx. Near Gifford's.
P. heterophylla. L. Common.
P. balsamifera. L. **Var. candicans.** Gray. Occasional.
P. alba. L. Common in cultivation.

CONIFERÆ.

Pinus. Tourn. **P. rigida.** Miller. Common.
P. inops. Ait. Tottenville ; also near Clifton.
P. mitis. Michx. One tree near Gifford's.
P. Strobus. L. Not uncommon.
Juniperus. L. **J. communis.** L. One tree of the erect variety grows in the " Cedars " near New Dorp.
J. Virginiana. L. Very common.

ARACEÆ.

Arisæma. Martius. **A. triphyllum.** Torrey. Common.
Peltandra. Raf. **P. Virginica.** Raf. Common.
Symplocarpus. Salisb. **S. fœtidus.** Salisb. Common.
Acorus. L. **A. Calamus.** L. Rather common.

LEMNACEÆ.

Lemna. L. **L. minor.** L. Common. Found in flower near Court House.
L. polyrrhiza. L. Common.
L. perpusilla. Torr. In ponds, south of the railroad track, - near Huguenot, etc.

TYPHACEÆ.

Typha. Tourn. **T. latifolia.** L. Common.
T. augustifolia. L. Frequent.
Sparganium. Tourn. **S. simplex.** Hudson. **Var. androcladum.** Gray. Common.

NAIADACEÆ.

Naias. L. **N. flexilis.** Rostk. Frequent.
Zostera. L. **Z. marina.** L. Common along the shore.
Ruppia. L. **R. maritima.** L. Salt ditch, near New Dorp.
Potamogeton. Tourn. **P. natans.** L. Common.
P. hybridus. Michx. "Silver Lake." (*W. H. L. in T. C. B.*)
P. perfoliatus. L. Not rare.
P. pusillus. L. Near Clove Lake.
P. pectinatus. L. Along salt water ditches.

ALISMACEÆ.

Triglochin. L. **T. maritimum.** Salt marshes ; rather common.
Alisma. L. **A. Plantago.** L. **Var. Americanum.** Gray. Common.
Sagittaria. L. **S. variabilis.** Engelm. Very common, with its varieties.

ORCHIDACEÆ.

Orchis. L. **O. spectabilis.** L. Not rare.
Habenaria. Willd. **H. tridentata.** Hook. Rare. Found once near Clove Lake.
H. virescens. Spreng. Common.
H. ciliaris. R. Br. Abundant in the Linden Park Swamp, and " near Prince's Bay." (*Leggett.*)
H. lacera. R. Br. Frequent.
H. psycodes. Gray. Rather common.
Goodyera. R. Br. **G. pubescens.** R. Br. Frequent.
Spiranthes. Richard. **S. cernua.** Richard. Common.
S. graminea. Lindl. **Var. Walteri.** Gray. Linden Park Swamp, also Tottenville.
S. gracilis. Bigelow. Common.
S. simplex. Gray. Very sparingly near Tottenville. .
Listera. R. Br. **L. cordata.** R. Br. (*Elliot Collection.*)
Pogonia. Juss. **P. ophioglossoides.** Nutt. Clove Lake Swamp.
P. pendula. Lindl. "On the high hills of S. I." (*Torr. Cat.*)
P. verticillata. Nutt. Sparingly in woods near Gifford's and Huguenot.
Calopogon. R. Br. **C. pulchellus.** R. Br. Clove Lake Swamp.
Tipularia. Nutt. **T. discolor.** Nutt. Rare ; woods near Egbertville.
Microstylis. Nutt. **M. ophioglossoides.** Nutt. "In a glen near New Dorp." (*Mr. A. Brown.*)
Liparis. Richard. **L. lillifolia.** Richard. Frequent.
L. Lœselii. Richard. "On Staten Island, in the gravelly bank of a railroad cutting." (*I. H. Hall in T. C. B.*)
Corallorhiza. Haller. **C. odontorhiza.** Nutt. Woods near Egbertville, and near Clove Lake.
C. multiflora. Nutt. Near Clifton. Ocean Terrace.
Cypripedium. L. **C. acaule.** Ait. Common.
C. spectabile. Swartz. "Near the Fort, Clifton." (*W. Chorlton.*)

AMARYLLIDACEÆ.

Hypoxys. L. **H. erecta.** L. Common.

HÆMODORACEÆ.

Aletris. L. **A. farinosa.** L. Rather common.

IRIDACEÆ.

Iris. L. **I. versicolor.** L. Common.
I. Virginica. L. Along salt meadows near New Dorp.
Sisyrinchium. L. **S. Bermudiana.** L. Very common.

DIOSCOREACEÆ.

Dioscorea. Plumier. **D. villosa.** L. Common. In fruit this plant becomes much more villous than in flower.

SMILACEÆ.

Smilax. Tourn. **S. rotundifolia.** L. Very common. **Var. guadrangularis.** Frequent.
S. glauca. Walt. Near Linden Park, along the beach. Tottenville.
S. herbacea. L. Common.

LILIACEÆ.

Trillium. L. **T. erectum.** L. (*Elliot Collection.*)
i **T. cernuum.** L. Frequent.
Medeola. Gronov. **M. Virginica.** L. Common.
Melanthium. Gronov. **M. Virginicum.** L. Common in the Clove Lake Swamp.
Veratrum. Tourn. **V. viride.** Ait. Rather common.
Chamælirium. Willd. **C. luteum.** Gray. Linden Park Swamp.
Uvularia. L. **U. perfoliata.** L. Common.
U. sessilifolia. L. Common.
Smilacina. Desf. **S. racemosa.** Desf. Common.
S. stellata. Desf. Sea Beach, near Clifton, Tottenville. Clove Lake Swamp.
S. bifolia. Ker. Very common.
Polygonatum. Tourn. **P. biflorum.** Ell. Common.
P. giganteum. Dietrich. Near Clove Lake Swamp.
Asparagus. L. **A. officinalis.** L. Frequent along salt meadows. Also in fields.
Lilium. L. **L. Philadelphicum.** L. Not rare.
L. Canadense. L. Rather common.
L. superbum. L. Common along salt meadows near New Dorp and Garretson's.
Erythronium. L. **E. Americanum.** Smith. Common.
Ornithogalum. Tourn. **O. umbellatum.** L. Very common.
Allium. L. **A. tricoccum.** Ait. Clove Lake Swamp, and near Springville.
A. vineale. L. Very common.
A. Canadense. Kalm. Common.
Hemerocallis. L. **H. fulva.** L. Frequent.

JUNCACEÆ.

Luzula. DC. **L. campestris.** DC. Common.
Juncus. L. **J. effusus.** L. Very common.
J. marginatus. Rostk. Common. **Var. paucicapitatus.** Frequent. **Var. biflorus.** "Rossville, S. I." (*T. C. B.*)

J. bufonius. L. Common.
J. Gerardi. Loisel. Common on salt marshes.
J. tenuis. Willd. Common.
J. Greenii. Oakes & Tuckerman, "S. I." (*W. H. L. in T. C. B.*)
J. articulatus. L. Frequent.
J. acuminatus. Michx. **Var. legitimus.** Common.
J. scirpoides. Lam. **Var. macrostemon.** "S. I." (*Austin in T. C. B.*) "Tottenville." (*W. H. L. in T. C. B.*)
J. Canadensis. J. Gay. Rather common.

PONTEDERIACEÆ.

Pontederia. L. **P. cordata.** L. Common.

XYRIDACEÆ.

Xyris. L. **X. flexuosa.** Muhl. Chapm. Pine woods near Tottenville.

CYPERACEÆ.

Cyperus. L. **C. flavescens.** L. Salt meadows near New Dorp.
C. diandrus. Torr. Common.
C. Nuttallii. Torr. Rather common.
C. dentatus. Torr. "S. I." (*W. H. L. in T. C. B.*)
C. phymatodes. Muhl. Near Court House and Richmond.
C. strigosus. L. Very common.
C. Michauxianus. Schultes. Frequent.
C. Grayii. Torr. Common on the beach.
C. filiculmis. Vahl. Common in sandy places.
C. ovularis. Torr. Rather common.
C. retrofractus. Torr. Pine woods near Tottenville, sparingly.
Dulichium. Richard. **D. spathaceum.** Pers. Common.
Eleocharis. R. Br. **E. obtusa.** Schultes. Common.
E. palustris. R. Br. Common.
E. tenuis. Schultes. Rather common.
E. acicularis. R. Br. Frequent.
E. pygmæa. Torr. Common on salt meadows.
Scirpus. L. **S. pungens.** Vahl. Common.
S. validus. Vahl. Common.
S. maritimus. L. Common on salt meadows.
S. polyphyllus. Vahl. Reported from S. I. in the Bulletin, but needs confirmation.
S. Eriophorum. Michx Common.
S. atrovirens. Muhl. Rather common.
Fimbristylis. Vahl. **F. autumnalis.** Rœm & Schultes. Common.
F. capillaris. Gray. Common.
F. spadicea. Vahl. **Var. castanea.** Gray. Salt meadows near Richmond.

Rhynchospora. Vahl. **R. glomerata.** Vahl. Not common. Near Garretson's.
Carex. L. **C. polytrichoides.** Muhl. Common.
C. vulpinoidea. Michx. Common.
C. stipata. Muhl. Common.
C. cephalophora. Muhl. Common.
C. rosea. Schk. Rather common.
C. stellulata. L. **Var. scirpoides.** Gray. Common.
C. scoparia. Schk. Common.
C. fœnea. Willd. Salt meadows. **Var. sabulonum.** Gray. Common on the beach.
C. straminea. Schk. Common.
C. stricta. Lam. Very common.
C. crinita. Lam. Common.
C. granularis. Muhl. Common.
C. grisea. Wahl. "Near Gifford's." (*J. W. Congdon.*)
C. virescens. Muhl. Rather common.
C. digitalis. Willd. Not rare.
C. laxiflora. Lam. Very common.
C. Pennsylvanica. Lam. Common.
C. varia. Muhl. Common.
C. miliacea. Muhl. Tottenville.
C. vestita. Willd. Tottenville.
C. comosa. Boott. Common.
C. hystricina. Willd. Rather common.
C. tentaculata. Muhl. Common.
C. intumescens. Rudge. Common.
C. lupulina. Muhl. Not rare.
C. folliculata. L. Common.
C. squarrosa. L. Not rare.
C. retrorsa. Schw. Common.
C. tenella. Schk. (*Mr. Ruger.*)

GRAMINEÆ.

Leersia. Solander. **L. Virginica.** Willd. Common.
L. oryzoides. Swartz. Common
Zizania. Gronov. **Z. aquatica.** L. "West Brighton." (*G. W. Wright.*)
Phleum. L. **P. pratense.** L. Common. A peculiar variety of this species is often found late in the season with all its flowers degenerated into bracts and leaflets. One field near Four Corners seemed to be particularly full of it.
Vilfa. Adans. Beauv. **V. aspera.** Beauv. Rather common.
V. vaginæflora. Torr. Common.
Agrostis. L. **A. perennans.** Tuckerman. Common.
A. scabra. Willd. Common.
A. vulgaris. With. Common.
Cinna. L. **C. arundinacea.** L. Not rare.
Muhlenbergia. Schreber. **M. Mexicana.** Trin. Common.
M. sylvatica. Torr and Gray. Not rare.
5

M. diffusa. Schreber. Very common.
Calamagrostis. Adans. **C. arenaria.** Roth. Common on the beach.
C. Nuttalliana. Steud. Near New Brighton.
Stipa. L. **S. avenacea.** L. Tottenville.
Aristida. L. **A. dichotoma.** Michx. Common.
A. gracilis. Ell. Common.
A. purpurascens. Poir. Todt Hill. Not rare.
A. tuberculosa. Nutt. Tottenville. Point near Gifford's.
Spartina. Schreber. **S. cynosuroides.** Willd. Salt meadows.
S. polystachya. Willd. Mulh. With the last.
S. juncea. Willd. Common on salt meadows.
S. stricta. Roth. **Var. glabra.** Gray. Salt marshes.
Eleusine. Gaertn. **E. Indica.** Gaertn. Common.
Leptochloa. Beauv. **L. fascicularis.** Gray. Sparingly in salt meadows near New Dorp.
Tricuspis. Beauv. **T. seslerioides.** Torr. Common.
T. purpurea. Gray. Common on the beach.
Dactylis. L. **D. glomerata.** L. Common.
Eatonia. Raf. **E. Pennsylvanica.** Gray. Common.
Glyceria. R. Br. Trin. **G. Canadensis.** Trin. Swamp near Gifford's.
G. elongata. Trin. Near Court House.
G. nervata. Trin. Common.
G. pallida. Trin. Rather common.
G. fluitans. R. Br. Common.
G. maritima. Wahl. Sea beach, not very common.
G. acutiflora. Torr. (*Mr. Ruger.*)
Brizopyrum. Link. **B. spicatum.** Hook. Common on salt meadows.
Poa. L. **P. annua.** L. Very common.
P. compressa. L. Common.
P. pratensis. L. Common.
P. trivialis. L. Near Gifford's.
Eragrostis. Beauv. **E. pectinacea.** Gray. Common.
E. pilosa. Beauv. Common around New Brighton.
Festuca. L. **F. tenella.** Willd. Rather common.
F. ovina. L. Very common.
F. elatior. L. Not very common.
Bromus. L. **B. racemosus.** L. New Dorp. Tottenville
B. secalinus. L. Very common.
B. ciliatus. L. Quite common.
Phragmites. Trin. **P. communis.** Trin. Common.
Lolium. L. **L. perenne.** L. Not rare.
Triticum. L. **T. repens.** L. Very common.
Hordeum. L. **H. jubatum.** L. Fort Hill. Rare.
Elymus. L. **E. Canadensis.** L. Not rare.
E. striatus. Willd. Near Egbertville.
Danthonia. DC. **D. spicata.** Beauv. Common.
Trisetum. Persoon. **T. palustre.** Torr. Near Garretson's, sparingly.

Aira. L. **A. flexuosa.** L. Not rare.
A. cæspitosa. L. Rather common.
Holcus. L. **H. lanatus.** L. Common.
Hierochloa. Gmelin. **H. borealis.** Rœm & Schultes. Not rare.
Anthoxanthum. L. **A. odoratum.** L. Common.
Phalaris. L. **P. arundinacea.** L. Not very common.
Paspalum. L. **P. setaceum.** Michx. Very common.
P. læve. Michx. Woods near Richmond.
Panicum. L. **P. filiforme.** L. Common.
P. glabrum. Gaudin. Quite common.
P. sanguinale. L. Very common.
P. agrostoides. Spreng. Frequent.
P. proliferum. Lam. Common.
P. capillare. L. Very common.
P. virgatum. L. Common.
P. amarum. Ell. Near Huguenot and Prince's Bay. Near New Dorp.
P. latifolium. L. Common.
P. clandestinum. L. Common.
P. microcarpon. Muhl. Sparingly in woods near Richmond.
P. pauciflorum. Ell. Near Richmond.
P. dichotomum. L. Very common.
P. depaupuratum. Muhl. Common.
P. verrucosum. Muhl. Near Tottenville.
P. Crus-galli. L. Very common.
Setaria. Beauv. **S. glauca.** Beauv. Common.
S. viridis. Beauv. Not rare.
Cenchrus. L. **C. tribuloides.** L. Common.
Tripsacum. L. **T. dactyloides.** L. Near Marsh land.
Andropogon. L. **A. furcatus.** Muhl. Not common.
A. scoparius. Michx. Very common.
A. Virginicus. L. Tottenville.
Sorghum. Pers. **S. nutans.** Gray. Common.

EQUISETACEÆ.

Equisetum. L. **E. arvense.** L. Very common.
E. hyemale. L. Tottenville, and near Four Corners.

[FILICES.

Polypodium. L. **P. vulgare.** L. "Near Four Corners." (*W. Chorlton.*) "Near Gifford's." (*J. J. Crooke.*)
Adiantum. L. **A. pedatum.** L. Common.
Pteris. L. **P. aquilina.** L. Very common.
Woodwardia. Smith. **W. Virginica.** Smith. Not rare.
W. angustifolia. Smith. Not rare.
Asplenium. L. **A. ebeneum.** Ait. Common.
A. thelypteroides. Michx. Common.
A. Filix-fœmina. Bernh. Common

Phegopteris. Féc. **P. hexagonoptera.** Féc. Common.
Aspidium. Swartz. **A. Thelypteris.** Swartz. Common.
A. Noveboracense. Swartz. Common.
A. spinulosum. Swartz. **Var. intermedium.** Gray. Very common.
A. cristatum. Swartz. Not common. **Var. Clintonianum.** Gray. Clove Lake Swamp. Sparingly.
A. marginale. Swartz. Frequent.
A. acrostichoides. Swartz. Very common. **Var. incisum** Gray. Woods near Egbertville.
Cystopteris. Bernh. **C. fragilis.** Bernh. Abundant in woods near Egbertville.
Onoclea. L. **O. sensibilis.** L. Very common.
Woodsia. R. Br. **W. obtusa** Torr. Near Egbertville.
Dicksonia. L. Her. **D. punctilobula.** Kunze. Common.
Osmunda. L. **O. regalis.** L. Common.
O. Claytoniana. L. Common.
O. cinnamomea. L. Very common.
Botrychium. Swartz. **B. Virginicum.** Swartz. Common.
B. ternatum. Var. obliquum. Gray. Common. **Var. dissectum.** Gray. Usually grows with the last.

LYCOPODIACEÆ.

Lycopodium. L. **L. lucidulum.** Michx. Common.
L. inundatum. L. **Var. Bigelovii.** Tuckerman. Sparingly near Tottenville.
L. dendroideum. Michx. Common.
L. clavatum. L. On Ocean Terrace. Not common.
L. complenatum. L. Common.
Selaginella. Beauv. **S. apus.** Spring. Common.
S. rupestris. Spring. Near Four Corners. Sparingly.

JANUARY, 1880.

LIST OF UNITED STATES
PHALLOIDEI.

Phallus L. (A. Hymenophallus *Nees.*)
    indusiatus Vent., S. Car., Pa. (Schw.), Mass. (Frost). Conn.
    (Eaton.)
    Dæmonum Rumph, Ohio (Lea), N. Y. (Peck).
    duplicatus Bosc. S Car., Pa. (Schw.), N. Y. (Ger.), Mass.
    (Frost: Sprague), Conn. (Wright.)
    Ravenelii B. & ., S. Car., (Rav.), N. Y. (Pk.)
    (B. Ithyphallus *Fr.*)
    impudicus L. S. Car., Schw.), N. Y. (Pk. Ger.) Mass. (Far-
    low : Frost.), Ohio (Lea,)
    (C. Leiophallus *Fr.*)
    rubicundus Bosc. S. Car.. N.Y. (Schw.), Mass. (Frost.)

Cynophallus Fr.
    caninus Schaeff. S Car., (Curt.) Mass. (Frost.) N.Y (Warne.)

Corynites B. & .C.
    Ravenelii B. & C. S. Car., (Curt.) N. Y. (Pk.: Howe : Ger.
    Curtisii Berk. Conn. (Wright.)
    brevis B. & C. (Inserted in Curtis' Catalogue but not
    mentioned in Berkeley's Notices of N. A. Fungi.)

Simblum Klotzsch.
    rubescens Ger. N. Y. (Trask ; Rodman ; Halsey.)

Clathrus Mich.
    cancellatus L. Ga. (Leconte, *fide* Schw.)

Laternea, Turp.
    columnata Bosc. S. ar. (Bosc.), Ga. (Leconte.)

§ 4. Flora of Richmond County, N. Y.
Additions and new localities, 1879.

Podophyllum peltatum L. Sparingly near Rossville.
Nasturtium palustre DC. Near Port Richmond and Woodrow.
Draba Caroliniana Walt. Near Rossville.
Sisymbrium Thalianum Gaud. Near Woodrow.
Camelina sativa Crantz. Rossville village.
Linum usitatissimum L. Sparingly at Clove Lake.
Coronilla varia L. Roadsides near New Dorp and Giffords.
Desmodium laevigatum DC. Sparingly near Tottenville.
Trifolium incarnatum L. Sparingly in waste ground near Rich-
    mond.
Rubus strigosus Michx. Near Pleasant Plains.
Fragaria Indica L. In Richmond Village.
Ribes floridum L. Near Pleasant Plains.
Coriandrum sativum L. Sparingly along the Sea Beach, near New
    Dorp.
Cicuta bulbifera L. Linden Park Swamp.
Lonicera parviflora Lam. Near Prince's Bay, 1870, [W. H.
    Leggett.]
Sericocarpus solidagineus Nees. Tottenville.
Aster concolor L. Kreischerviile.
Solidago patula Muhl. Clove Lake Swamp.

Solidago neglecta T. & G. Clove Lake Swamp.

Vaccinium Pennsyvanicum Lam. Rather abundant about Tottenville.

Vaccinium macrocarpon Ait. Sparingly near Clove Lake. Near Richmond.

Gaylussacia frondosa T. & G. On Todt Hill, near New Dorp, and at Tottenville. Frequent.

Monarda fistulosa L. On Richmond Hill, and near Willow Brook.

Lophanthus nepetoides Benth. New Dorp.

Phlox subulata L. Near Rossville. (G. W. Wright.)

Solanum rostratum Dunal. A single plant near Four Corners, 1875, (W. H. Rudkin.)

Artemisia vulgaris L. Port Richmond.

Atriplex patula L. var. **littoralis** Gray. West New Brighton.

Quercus Phellos L. Sparingly at Tottenville.

Sparganium eurycarpum Engl. Near Garretsons.

Vallisneria spiralis L. Clove Lake Swamp.

Calopogon pulchellus R Br. Linden Park Swamp,

Habenaria ciliaris Lindl. Tottenville.

Habenaria tridentata Hook. Tottenville.

Tradescantia Virginica L. Escaped from gardens at Tottenville.

Cyperus cylindricus N. L. Britton. Tottenville.

Eriophorum Virginicum L. Near Richmond.

Rhynchospora alba Vahl. Clove Lake Swamp.

Cladium mariscoides Torr. Linden Park Swamp.

Carex subulata Michx. Swamps near the railroad. (W. H. Leggett.)

Carex monile Tuck. Near New Dorp.

Carex lanuginosa Michx. Near New Dorp.

Carex sterilis Willd. Rather common.

Agrostis alba L. Common.

Calamagrostis Canadensis Beauv. Linden Park, and Clove Lake Swamps.

Calamagrostis Nuttalliana Steud. Watchogue and Tottenville.

Eatonia obtusata Gray. Linden Park Swamp.

Glyceria obtusa Trin. Tottenville.

Glyceria acutiflora Torr. Near Bull's Head.

Eragrostis poaeoides Beauv. var. **megastachya** Gray. Port Richmond.

Elymus Virginicus, L. Tottenville.

Uniola gracilis Michx. Tottenville.

Arrhenatherum avenaceum Beauv. Clifton, and Richmond Village.

Andropogon macrourus Michx. Tottenville.

CORRECTION.

Cyperus retrofractus Torr. (?) should be **C. cylindricus** N. L. Britton. Sparingly at Tottenville.

C. A. Hollick,

N. L. Britton.

Terms One dollar per annum beginning with the January No. Address, P. V. LeRoy, Herbarium, Columbia College, 49th St. and Madison Ave., N. Y. Money orders on Station H, N. Y.

Communications for the editor should be addressed to Wm. H. Leggett, 54, East 81st Street, New York.

Back Volumes and copies of the Constitution and By-Laws of the Club may be obtained of the editor. Money orders on Station K, N. Y.

The Club meets regularly the second Tuesday of the month in the Herbarium, Columbia College, at 7:30 P. M. Botanists are invited to attend.

Ranunculus aquatilis, L., var. *trichophyllus*, Chaix. Abundant in
Clove Lake Swamp; has appeared spontaneously since last year.

Dentaria laciniata, Muhl. Woods near New Dorp. (Miss C. O.
Thompson).

Brassica campestris, L., West New Brighton. Introduced in ballast.
Also occasional in old fields.

Brassica rapa, L. Occasionally spontaneous in cultivated fields.

Ascyrum Crux-Andreae, L. Kreischerville.

Vaccaria vulgaris, Host. Tompkinsville. (Miss C. O. Thompson.)

Silene inflata, Smith. New Dorp. Rare.

Stellaria longipes, Goldie. Abundant near Port Richmond.

Stellaria uliginosa, Murr. Rossville. Rare.

Sagina decumbens, T. & G. Roadsides near Woodrow.

Sagina decumbens, T. & G., var. *Smithii*, Gray. Garretson's. Rare.

Gypsophila arvensis, L. New Brighton, in gardens and waste-places.
Rare. Introduced.

Malva moschata, L. A single plant near Court House. 1880.

Zanthoxylum Americanum, Mill. Port Richmond.

Rhus typhina, L. Richmond Hill.

Medicago sativa, L. West New Brighton. Introduced in ballast.

Desmodium Canadense, DC. Clove Lake.

Tephrosia Virginiana, Pers. Common around Tottenville.

Prunus Mahaleb, Linn. Escaped to roadsides near Garretson's.
Rare.

Prunus Cerasus, L. Thoroughly established in woods and copses,
probably through the agency of birds.

Pirus communis, L. Sparingly established in woods and borders of
fields.

Amygdalus Persica, L. Along old stone walls and open woods.

Amelanchier Canadensis, T. & G., var. *rotundifolia*, Gray. Prince's
Bay.

Sedum ternatum, Michx. Roadsides near Woodrow. Rare.

Aethusa Cynapium, L. Near Clove Lake. Rare.

Carum Carui, L. Roadsides near Four Corners.

Symphoricarpus vulgaris, Michx. Roadsides and waste-places near
gardens (escaped).

Lonicera Tartarica, L. Roadside near Huguenot; escaped from
gardens.

Viburnum nudum, L., var. *cassinoides*, Gr. Swamps near Watchogue.

Galium trifidum, L., var. *pusillum*, Gr. Near Pleasant Plains. Rare.

Oldenlandia glomerata, Michx. Not rare at Tottenville.

Fedia olitoria, Vahl. Streets of West New Brighton. Rare. (Miss
Webster.)

Eupatorium album, L. Common at Tottenville.

Solidago puberula, Nutt. Not rare at Watchogue.

Matricaria Chamomilla, L. West New Brighton. Introduced in
ballast.

Artemisia biennis, Willd Rather plentiful near Clove Lake.

Serratula tinctoria, L. West New Brighton. Probably introduced
in ballast.

Senecio aureus, L., var. *obovatus*, Gray. Near Richmond.

Helianthus strumosus, L. Near Clove Lake.

Helianthus decapetalus, L. Near Clove Lake.

Lysimachia thyrsiflora, L. Linden Park Swamp.

Mimulus alatus, Ait. Along a brook near Huguenot.

Galeopsis Ladanum, L. "Staten Island." BULL. TORR. BOT. CLUB, vi., p. 115.

Perilla ocimoides, L,. var. *crispa*, Gray. "Escaped to roadsides near Huguenot." (W. H. Leggett, in BULL. TORR. BOT. CLUB, ii., 29.)

Echinospermum Lappula, Lehm. West New Brighton. Introduced in ballast.

Cynoglossum officinale, L. One plant near Concord. 1880.

Cuscuta compacta, Juss. Tottenville.

Solanum Carolinense, L. Waste-places and roadsides, New Dorp and New Brighton.

Physalis angulata, L. New Dorp. Abundant in one field.

Nicotiana rustica, L. One plant near Clove Lake. 1880.

Aristolochia Serpentaria, L. Tottenville. (Herb. W. H. Leggett.)

Rumex orbiculatus, Gray. Clove Lake Swamp.

Callitriche Austinii, Engelm. Tottenville and top of Ocean Terrace.

Croton capitatus, Michx. Streets of New Brighton. (Miss C. O. Thompson.)

Cannabis sativa, L. West New Brighton. Common. Introduced in ballast.

Potamogeton pauciflorus, Pursh. "Near Quarantine." (W. H. Leggett).

Sisyrinchium Bermudiana, L., var. *mucronatum*, Gray. Tottenville.

Smilax tamnoides, L. "Staten Island, one plant only." (T. F. Allen. 18th Annual Rept. State Cab. Nat. Hist., p. 202.)

Helonias bullata, L. In a maple swamp near Kreischerville.

Eleocharis melanocarpa, Torr. New Springville and New Dorp.

Scirpus planifolius, Muhl. Tottenville and Garretson's.

Scirpus sylvaticus, L. Near Garretson's.

Scirpus Eriophorum, Michx., var. *laxus*, Gr. Near Richmond.

Carex triceps, Michx. New Dorp.

Carex Muhlenbergii, Schk , var. *enervis*, Boott. (W H. Leggett, in BULL. TORR. BOT. CLUB, vi., 9.)

Carex rosea, Schk., var. *radiata*, Dew. Near Garretson's.

Carex utriculata, Boott. Linden Park Swamp.

Carex canescens, L. Common.

Agrostis spica-venti, L. West New Brighton. Introduced in ballast.

Phalaris arundinacea, L., var. *picta*, Gr. Thoroughly established along a brook near Four Corners.

Poa nemoralis, L West New Brighton. Introduced in ballast.

Poa serotina, Erhr. West New Brighton. Introduced in ballast.

Festuca nutans, Willd. Hills back of Garretson's. Not common.

Festuca ovina, L., var. *duriuscula*, Gray. Near Garretson's.

Bromus multiflorus, Smith. West New Brighton. Introduced in ballast.

Bromus mollis, L. New Dorp. Not common.

Panicum Crus-galli, L., var. *hispidum*, Gr. Common on the salt marshes.

Polypodium vulgare, L. Sparingly near Silver Lake, Four Corners, Richmond, and top of Ocean Terrace.

ARTHUR HOLLICK,
N. L. BRITTON.

Flora of Richmond Co., N. Y.—Additions, corrections and new localities, 1883–1884. ᴄ | eᴜ ᴄ ᴋ ᴄ ᳕Ꞓᴏ. ᳚ᴢ

Delphinium Consolida, L.—Roadsides near Richmond.

Adonis autumnalis, L.—Stapleton Flats; introduced in ballast. (Miss C. O. Thompson.)

Thlaspi arvense, L.—Clove Lake.

Reseda odorata, L.—"Waste places, S. I., 1865;" (Wm. H. Leggett.) Mr. Samuel Henshaw, also, states that he knew of one plant which had grown and flourished for several successive years in the crevice of a stone wall in Stapleton.

Viola pubescens, Ait., var. *eriocarpa*, Nutt. Near Barrett's Dye Works, Port Richmond.

Ascyrum Crux-Andreæ, L.—Richmond Valley and Tottenville.

Vaccaria vulgaris, Host.—Stapleton Flats; introduced in ballast. (Miss C. O. Thompson.)

Silene nocturna, L.—Stapleton Flats; introduced in ballast. (Miss C. O. Thompson.)

Stellaria graminea, L. Tottenville, Four Corners, New Brighton and New Springville. Appears to be spreading and is much less rare than formerly. (Replaces *S. borealis*, Bigel., of our catalogue.)

Vicia hirsuta, Koch.—New Brighton. Rare.

Vitis cordifolia, Mchx , var. *riparia*, Gr.—Garretsons.

Cytissus triflorus, L'Her. Todt Hill; escaped from cultivation.

Prunus Americana, Marshall. Abundant in a limited locality near the Moravian Cemetery.

Rubus laciniatus, Willd.—Silver Lake ; escaped from cultivation.

Pirus arbutifolia, L., var. *erythrocarpa*, Gr.—Common.

Pirus arbutifolia, L., var. *melanocarpa*, Gr.—Common. (This and the preceding variety should replace *P. arbutifolia*, L., in our catalogue.)

Lonicera Japonica, Thunb.—Woods and roadsides near Four Corners and Richmond. Rapidly spreading, from pieces thrown out of gardens.

Viburnum Lentago, L.—Clove Lake Swamp.

Galium trifidum, L., var. *latifolium*, Torr.—"Clifton, S. I., 1870." (Wm. H. Leggett.)

Galium verum, L.—Stapleton Flats and New Brighton. (Miss C. O. Thompson.)

Galium Mollugo, L.—New Brighton. (Miss C. O. Thompson.)

Asperula arvensis, L.—Stapleton; introduced in ballast. (Miss C. O. Thompson.)

Aster multiflorus, Ait.—Abundant on the salt marshes near Green Ridge.

Aster Novæ-Angliæ, L., var. *roseus*, Gr.—Port Richmond and West New Brighton.

Solidago Canadensis, L., var. *scabra*, T. & G.—Tottenville.

Eclipta alba, Hassk.—Beginning to spread in waste places and gardens. (*E. procumbens*, Michx., of our catalogue.)

Chrysanthemum Parthenium, Pers.—"Rubbish heaps, S. I. 1869." (Wm. H. Leggett.) New Brighton. (Miss C. O. Thompson.)

Centaurea solstitialis, L.—Stapleton Flats; introduced in ballast. (Miss C. O. Thompson.) "Stapleton, S. I., 1870." (Wm. H. Leggett.) (Replaces *C. Militensis*, L., of our catalogue.)

Leontodon autumnale, L.—New Brighton. (Dr. F. Hollick.)

Mulgedium Floridanum, D. C. Todt Hill.

Gaultheria procumbens, L.—Watchogue. (E. M. Eadie.)

Campanula rapunculoides, L.—Roadsides, Richmond Valley.

Primula veris, L.—" Roadside, West New Brighton, 1879." (G. M. Wilber.)

Paulownia imperialis, Sieb. and Zucc.—Silver Lake; spreading quite extensively.

Trichostema lineare, Nutt. Tottenville,

Pycnanthemum Torreyi, Benth. "Richmond Valley, 1864." (Wm. H. Leggett.)

Monarda didyma, L. Willow Brook; escaped from gardens.

Physostegia Virginiana, Benth. "Staten Island, 1861." (Wm. H. Leggett.)

Marrubium vulgare, L.—Watchogue. (E. M. Eadie.)

Echium vulgare, L.—Kreischerville. (Wm. T. Davis.)

Asperugo procumbens, L. Stapleton Flats; introduced in ballast. (Miss C. O. Thompson.)

Ipomœa Nil, Roth.—New Brighton. Rare.

Nicandra physaloides, Gærtn,—Abundant at Tottenville.

Sabbatia chloroides, Pursh.—Chelsea. (E M. Eadie.)

Gentiana Saponaria, L. Tottenville.

Asclepias variegata, L. Tottenville. (Wm. H. Rudkin.)

Atriplex rosea, L.—Waste places, Clifton.

Polygonum incarnatum, Ell.—New Brighton. (Miss C. O. Thompson.) Not, as first supposed, common.

Rumex pulcher, L.—Stapleton Flats ; introduced in ballast. (Miss C. O. Thompson.)

Ulmus fulva, Michx.—Egbertville.

Quercus bicolor. Willd.—Green Ridge and Egbertville.

Quercus prinoides, Willd.—Watchogue.

Betula nigra, L.—A few trees near Bull's Head.

Sparganium simplex, Huds., var *Nuttallii*, Engelm.—"Tottenville, 1871." (Wm. H. Leggett.)

Potamogeton pulcher, Tuck.—Silver Lake.

Potamogeton amplifolius, Tuck.—Clove Lake.

Sagittaria variabilis, Engelm., var. *obtusa*, Gr.—Common.

Sagittaria variabilis, Engelm., var. *hastata*, Gr.—Common. (These varieties should replace *S. variabilis*, Engelm., in our catalogue.)

Habenaria ciliaris, R. Br.—Watchogue. (E. M. Eadie.)

Smilax Pseudo China, L.—Linden Park Swamp.

Juncus dichotomus, Ell.—"Tottenville, 1871." (Wm. H. Leggett.)

Heleocharis prolifera, Torr. (?)—Found only in one clear deep spring, at Port Richmond. Has never been found in fruit, and hence the determination may prove to be wrong.

Heleocharis olivacea, Torr.—" Tottenville." (Wm. H. Leggett.)

Scleria triglomerata, Michx.—Linden Park.

Carex rosea, Schk., var. *radiata*, Desv.—"Huguenot, 1871." (Wm. H. Leggett)

Carex Muhlenbergii, Schk., var. *enervis*, Boot.—New Dorp and Garretsons.

Carex debilis, Michx.—" Rossville, 1869." (Wm. H. Leggett.)

Eragrostis pœoides, Beauv.—Streets of Port Richmond.

Eragrostis Purshii, Schrader.—New Brighton and Court House.

Phalaris Canariensis, L.—New Dorp and New Brighton.

Isoetes echinospora, Durieu, var. *Braunii*, Engelm. (?)—Only one specimen, found near Huguenot; not in fruit, the determination may therefore not be correct.

N. L. BRITTON.
ARTHUR HOLLICK.

Appendix No. 4

Flora of Richmond County, N. Y. Additions and New Localities, 1885.

Podophyllum peltatum, L. Pleasant Plains, (Miss Rich.)

Berberis vulgaris, L. Abundant at Tottenville.

Caulophyllum thalictroides, Michx. New Springville.

Nymphæa odorata, Ait., var. *minor*, Sims. Four Corners.

Papaver dubium, L. New Dorp. (Miss E. G. Knight.)

Papaver somniferum, L. Port Richmond.

Nasturtium palustre, DC. Stapleton Flats. (Miss C. O. Thompson.)

Erysimum cheiranthoides, L. In a field near Kreischerville.

Erysimum orientale, Br. Stapleton Flats. (Miss Thompson.)

Hesperis matronalis, L. Escaped near the Poor House.

Rapistrum rugosum, L. Stapleton Flats. (Miss Thompson.)

Lychnis vespertina, Sibth. New Brighton. (Miss Thompson.)

Rhus typhina, L. Tottenville, (W. T. Davis), Princes Bay.

Trifolium hybridum, L. New Dorp. (Miss E. G. Knight.)

Vicia Cracca, L. Richmond Village.

Lathyrus paluster, L. Garretsons, also a form intermediate between this and the var. *myrtifolius*, Gray.

Lespedeza reticulata, Pers. Frequent.

Prunus Pennsylvanica, L. Silver Lake. (Wm. T. Davis.)

Drosera rotundifolia, L. Tottenville.

Callitriche heterophylla, Pursh. Huguenot.

Scabiosa arvensis, L. Field near Richmond. (Miss Knight.)

Dipsacus sylvestris, Mill. Brickyards, Green Ridge.

Lonicera ciliata, Muhl. Hills back of Garretsons.

Coreopsis discoidea, Torr. and Gray. Huguenot.

Heliopsis bupthalmoides, Dunal. (?) Tompkinsville. (Miss C. O. Thompson.)

Lactuca Scariola, L. New Brighton. (Miss Thompson.)

Centaurea Cyanus, L. New Brighton. (Dr. F. Hollick.)

Hieracium aurantiacum, L. New Dorp and Pleasant Plains.

Hieracium Marianum, Willd. Frequent.

Vaccinium corymbosum, L., var. *glabrum*, Gray, Staten Island, (W. H. Leggett); var. *amoenum*, Gray, Giffords; var. *atrococcum*, Gray, Huguenot.

Campanula rapunculoides, L. Streets of New Brighton.

Verbascum Lychnitis, L. New Brighton. (Wm. Chorlton.)

Echium vulgare, L. Abundant in a field near the Poor House.

Lycopsis arvensis, L. In ballast at Stapleton. (Miss Thompson.)

Petunia nyctaginifolia. New Dorp and New Brighton, escaped.

Utricularia gibba, L. Tottenville. (W. H. Leggett.)

Lycopus Europæus, L. Clove Lake and tributaries.

Thymus vulgaris, L. Clifton, 1864. (W. H. Leggett.)

Salix fragilis, L. New Dorp and Princes Bay, commonly bearing branched catkins, (See Bulletin, vol. vi., p. 312)

Salix cordata, Muhl., var. *angustata*, Gray. Tottenville.

Smilax tamnoides, L. Sparingly near Kreischerville.

Potamogeton pusillus, L., var. *tenuissimus*, Fries. Clove Lake.

Muscari racemosum, Mill. Huguenot and New Brighton.

Chamælirium Carolinianum, Willd. Clove Lake.

Juncus acuminatus, Michx., var. *debilis*, Englm. New Dorp.

Cyperus diandrus, Torr., var. *castaneus*, Torr. Frequent.

Carex lagopodiodes, Schk. Staten Island. (W. H. Leggett.)

Carex laxiflora, Lam., var. *blanda*, Boott. Garretsons; var. *plantaginea*, Boott., near Clove Lake.

Eatonia Dudleyi, Vasey, incd. Giffords, (W. H. Leggett) ; near Garretsons.

Panicum microcarpum, Muhl. Court House.

Panicum nitidum, Lam. Frequent, also a hairy form answering the description of *P. pubescens*, Lam.

Panicum discolor, Chapm. Todt Hill and Tottenville, (Named by Mr. F. L. Scribner.)

Setaria Italica, Kunth. Roadsides near Richmond.

Botrychium ternatum, Swartz. sub-var. *intermedium*, Eaton. Tottenville and Garretsons.

Lycopodium complanatum, L., var. *sabinæfolium*, Gray. Ocean Terrace.

Three former lists of additions to our Catalogue of Staten Island Plants have been published in this Bulletin (vol. vii., pp. 11 12 ; vol. viii., p. 48 ; vol. ix., pp. 149-151 ; and vol. xii., pp. 38-40; these have all been reprinted and will be furnished to those desiring them on application to either of the undersigned.

ARTHUR HOLLICK.

N. L. BRITTON.

132

᾿Ὰ ·   ,¹ ᴼ ᴼᶜ

(Reprinted from BULLETIN OF THE TORREY BOTANICAL CLUB, Vol. XVI., No. 5)

Flora of Richmond Co., N. Y.—Additions and New Localities, 1886-1889.

APPENDIX No. 5.

Clematis ochroleuca, Ait.   Sand dune on the borders of salt meadows, near Watchogue.   A number of plants of the lobed leaved form with it.   (Wm. T. Davis.)

Ranunculus septentrionalis, Poir.   Clove Lake swamp.   A remarkable tendency to fasciation has been found in these specimens.

Nasturtium sylvestre (L.), R. Br.   Woods of Arden, near the shore.   (Mrs. N. L. Britton.)

Lechea racemulosa, Lam.   "Tottenville," *fide* specimens in Herb. W. H. Leggett.

Drosera intermedia, Drev. & Hayne, var. *Americana*, DC. Clove Lake swamp.

Malva sylvestris, L.   New Brighton.   Escaped from gardens.   (Wm. T. Davis.)

Trifolium hybridum, L.   New Brighton.   Becoming common.

Lathyrus maritimus (L.), Bigel.   New Dorp.

* *Rosa humilis*, Marshall, var. *lucida* (Ehrh.), Best. (?) "Stat. Is. west side, July 22, 1869."   (W. H. Leggett.)

Cratægus coccinea, L.   Tottenville.   (Wm. T. Davis.)

Tiedemannia rigida (L.), Coulter & Rose, var. *longifolia* (Pursh), B. S. P.   Garretsons.

Lonicera xylosteum, L.   Admitted into the 4th appendix under the name *L. ciliata*, Muhl., from flowering specimens obtained from a single bush near Garretsons.   The same plant has lately been found by Mr. Wm. T. Davis, at New Brighton.   It appears to be thoroughly naturalized at both localities, and probably grew from seeds transported by birds.

Aster spectabilis, Ait.   Mariners' Harbor.

Aster cordifolius, L., var. *glabratus*, Porter.   Frequent.

Aster cordifolius, L., var. *lanceolatus*, T. C. P.   Egbertville.†

---

* *Rosa humilis*, Marshall—and *Rosa humilis*, Marshall, var. *villosa*, Best, replace the *R. lucida*, Ehrh. of our catalogue.

(† MEM.—The *Aster*, admitted into our original catalogue under the title *A. sagittifolius*, Willd., has been determined by Prof. T. C. Porter to be a form of *A. cordifolius*, L., and the former species must therefore be omitted from the list.)

Aster Novi-Belgii, L., var. *lævigatus* (T. & G.), Gray. Ocean Terrace. Rare.

Aster Novi-Belgii, L., var. *elodes* (T. & G.), Gray. Garretsons. In swamps along salt meadows.

Solidago patula, Muhl. Mariners' Harbor.

Heliopsis helianthoides (L.), B. S. P. Kreischerville.

Helianthus grosse-serratus, Martens. ? Green Ridge.

Lactuca hirsuta, Muhl. New Dorp.

Gaultheria procumbens, L. Eltingville.

Pyrola secunda, L. Richmond.

Pycnanthemum incanum (L), Michx. Ocean Terrace.

Lophanthus nepetoides (L.), Benth. Tottenville.

Cynoglossum officinale, L. Richmond. Rare. One plant near Concord, 1880. One plant near Richmond, 1888.

Sabbatia dodecandra (L.), B. S. P. Kreischerville. Abundant in salt meadow.

Amarantus hybridus, L. Streets of New Brighton.

Juglans cinerea, L. South shore of Staten Island. (Samuel Ackerly, in Trans. N. Y. State Agric. Soc., 1843, under name of *J. cathartica*, Michx.) Recently reported by W. T. Davis as abundant along Sandy Brook in Westfield. (See Proc. Nat. Sci. Assoc. S. I., April, 1889.)

Hicoria alba (Nutt.), Britton, var. *maxima* (Nutt.), Britton. Court House.

Quercus Phellos, L. Tottenville. (Wm. T. Davis.) A number of trees have been discovered since the original find noted in the 1st Appendix for 1879.

Quercus ilicifolia, Wang. Watchogue.

Quercus Rudkinii, Britton. Tottenville. (Wm. T. Davis.)

Quercus heterophylla, Michx. f. Tottenville. (Wm. T. Davis.)

Betula nigra, L. Very rare. Since the original find near Bulls Head, Mr. Wm. T. Davis has noted one tree near Old Place, two at Tottenville and four at Richmond—all young.

Salix candida, Willd. Garretsons.

Salix tristis, Ait. Tottenville.

Salix purpurea, L. Garretsons, Woodrow and Old Place. Probably originated from cuttings of cultivated trees thrown aside in brush heaps.

Tsuga Canadensis (L.), Carr. Old Place. One tree.

Zannichellia palustris, L. " Staten Island." (Flora of New York).

Potamogeton pauciflorus, Pursh. Woods of Arden.

Potamogeton pulcher, Tuck. Woods of Arden.

Microstylis unifolia, Michx. Egbertville. (Mrs. N. L. Britton.) (The second time that a single plant has been discovered.)

Smilax glauca, Walt. A form of this species grows abundantly on the sand near Mariners' Harbor, Tottenville and Kreischerville, which is apparently *S. spinulosa*, Smith.

Chamælirium luteum (L.), Gray. Court House. (K. B. Newell.)

Juncus Balticus, Dethard, var. *littoralis*, Engelm. New Dorp.

Juncus dichotomus, Ell. Mariners' Harbor.

Eleocharis tuberculosa (Michx.), R. Br. Mariners' Harbor.

Carex glaucodea, Tuckerm. Fields, Court House.

Panicum latifolium, L., var. *molle*, Vasey. New Dorp.

Panicum nitidum, Lam., var. *ramulosum* (Michx.), Vasey. Frequent or occasional.

Cystopteris fragilis (L.), Sw., var. *dentata*, Hook. Near Egbertville. This was wrongly determined to be the typical form and as such was admitted into the original catalogue. A few specimens only of the type have been found, near Martling's Pond.

Onoclea sensibilis, L., var. *obtusilobata*, Torr. New Dorp. (Mrs. N. L. Britton.)

Azolla Caroliniana, Willd. Naturalized in pools in the Clove Valley.

ARTHUR HOLLICK,
N. L. BRITTON.

From Bull. Torrey Bot. Club for XVII pub.

Flora of Richmond Co., N. Y.--Additions and New Localities, 1890.

Appendix No. 6.

Ranunculus Ficaria, L. Willow Brook. Scarce.

Ranunculus lacustris, Beck & Tracy. (*Ranunculus multifidus,* Pursh). In a single pond hole, Ocean Terrace.

Ranunculus septentrionalis, Poir., replaces *R. fascicularis* of our Catalogue.

Hesperis matronalis, L. Annadale.

Lechea racemulosa, Lam., listed as occurring at Tottenville, proves to be an error in determination.

V. blanda, Willd. var. *amœna,* (Le Conte), B.S.P. Clove Valley.

Hypericum Canadense, L. *var. majus,* Gray. Garretsons. (Mrs. N. L. Britton).

Tilia Americana, L. Near Richmond. (Wm. T. Davis).

Nemopanthes mucronata, (L.), Trel. Giffords. (Wm. T. Davis).

Euonymus Europæus, L. Tottenville.

Fragaria Indica, Andr. Garretsons. (Miss Timmerman).

Eupatorium hyssopifolium, L. Pleasant Plains.

Eupatorium perfoliatum, L., var. *truncatum,* Gray. Oakwood

Aster Radula, Ait. Mariners Harbor. (R. G. Eccles).

Aster vimineus, Lam. Garretsons.

Erigeron ramosus (Walt.), B.S.P., var. *discoideus* (Robbins) B.S.P. New Dorp.

Hieracium aurantiacum, L. Rossville.

Lactuca Canadensis, L. A peculiar form with linear, entire leaves. New Dorp.

Oxycoccus macrocarpus, Pers. Annadale. (Wm. T. Davis).

Veronica Chamaedrys, L. Princes Bay. (Mrs. N. L. Britton).

Plantago major, L., var. *minima* (DC.), Decne. Richmond Valley.

Alnus glutinosa, Willd. Todt Hill.

Salix purpurea, L. Rossville.

Populus heterophylla, L. Near Green Ridge. (Wm. T. Davis).

Microstylis unifolia (Michx.), B.S.P. Ocean Terrace. (Miss C. A. Timmerman). Abundant near Egbertville.

Liparis Loeselii, (L.) Rich. Garretsons. (Miss C. A. Timmerman).

Habenaria blephariglottis (Willd.), Torr. Mariners' Harbor. (R. G. Eccles).

Habenaria ciliaris, (L.) R. Br. Old Place. (Wm. T. Davis.) Arlington. (Dr. R. G. Eccles.) Bogardus Corners.

Belamcanda Chinensis (L.), Red. Tottenville.

Tradescantia Virginica, L. Bogardus Corners.

Lemna trisulca, L. Clove Valley. (Thos. Craig).

Eleocharis palustris (L.). R. Br. var. *glaucescens* (Willd.) Gray. Common.

Scirpus Olneyi, Gray. New Dorp.

Panicum miliaceum, L. Todt Hill.

Glyceria distans (L.), Wahl. New Dorp. ARTHUR HOLLICK.
N. L. BRITTON.

[Reprinted from THE BULLETIN OF THE TORREY BOTANICAL CLUB, Vol. 22, No. 11, Nov., 1895.]

Flora of Richmond Co., N. Y. Additions and new Localities, 1891-1895.

APPENDIX, No. 7.

Ranunculus Pennsylvanicus L. New Dorp, A. A. Tyler.

Aquilegia Canadensis L. Tottenville and Richmond Valley, Wm. T. Davis and G. H. Pepper.

Nymphaea rubrodisca (Morong) Greene. Bull's Head, J. V. Leng.

Roripa sylvestris (L.) Bess. Sailors' Snug Harbor, Dr. F. Hollick.

Silene nutans L. Arrochar. W. C. Kerr.

Silene vulgaris (Moench) Garcke. Tottenvllle. W. T. Davis.

Tilia Americana L. Willow Brook and New Springville, W. T. Davis.

Acer Saccharum Marsh. Moravian Cemetery and New Springville. Wm. T. Davis.

Acer platanoides L. New Brighton and Todt Hill, Wm. T. Davis.

Acer Negundo L. Port Richmond.

Medicago sativa L. New Brighton, W. T. Davis; Princes' Bay.

Coronilla varia L. Tottenville, W. T. Davis.

Ulex Europaeus L. Ward's Hill, Tompkinsville, Dr. F. Hollick. (A single plant which has persisted for several years.)

Amorpha fruticosa L. Egbertville, W. T. Davis.

Phaseolus polystachyus (L.) B.S.P. Egbertville, W. T. Davis.

Opulaster opulifolius (L.) Kuntze. Todt Hill, W. T. Davis; Willow Brook (some undoubtedly seedlings).

Rubus odoratus L. Todt Hill, W. T. Davis.

Crataegus coccinea L. Karle's Neck, W. T. Davis.

Crataegus punctata Jacq. West New Brighton.

Agrimonia mollis (T. & G.) Britton, and *Agrimonia striata* Michx., replace *A. Eupatoria* L. in our catalogue.

Valeriana officinalis L. Gifford's Lane (replaces *V. sylvatica* Banks, in our catalogue).

Onagra Oakesiana (Gray) Britton. New Dorp.

AEthusa Cynnapium L. Streets of New Brighton, W. T. Davis.

Anthriscus vulgaris (L.) Hoffm. New Dorp.

Philadelphus coronarius L.   Todt Hill, W. T. Davis (apparently established from old garden waste).

Solidago Elliottii T. & G.   Garrettsons and New Dorp.

Solidago patula Muhl.   Garrettsons and Richmond, W. T. Davis.

Gnaphalium purpureum L.   Egbertville.

Tussilago Farfara L.   Garrettsons.

Liatris spicata (L.) Nutt.   Mariners' Harbor, W. T. Davis.

Sericocarpus linifolius (L.) B.S.P.   Watchogue, W. T. Davis.

Centaurea nigra L.   Moravian Cemetery. ·

Onopordon Acanthium L.   Sailors' Snug Harbor, Dr. F. Hollick.

Schollera macrocarpa (Ait.) Britton.   Tottenville, W. T. Davis ; New Dorp, Mrs. N. L. Britton ; Kreischerville.

Azalea viscosa rosea Hollick.   Arlington, W. T. Davis.

Gaultheria procumbens L.   Giffords.

Kalmia angustifolia L.   Bogardus' Corners.

Pyrola secunda L.   Bogardus' Corners.

Fraxinus viridis Michx. f.   Clifton and New Dorp.

Cynoglossum officinale L.   Arlington, W. T. Davis.

Convolvulus Sepium repens (L.) Gray.   Oakwood.

Ipomoea pandurata (L.) Meyer.   Mariners' Harbor, W. T. Davis.

Gerardia purpurea paupercula Gray.   New Dorp.

Veronica Anagallis-aquatica L.   New Brighton, Dr. F. Hollick.

Pentstemon Digitalis (Sweet) Nutt.   New Dorp, Mrs. N. L. Britton ; West New Brighton, T. C. Leng.

Stachys cordata Ridd.   Eltingville, W. T. Davis.

Mentha gentilis L.   Egbertville, A. A. Tyler.

Mentha sativa L.   New Dorp.

Mentha citrata Ehrh.   Richmond Valley.

Conopholis Americana (L. F.) Wallr.   Todt Hill, Mr. Stottler.

Plantago aristata Michx.   St. George, in recently filled-in ground.

Brousonnetia papyrifera Vent.   West New Brighton, Mariners' Harbor and Richmond, W. T. Davis.

Alnus glutinosa Willd.   Egbertville, W. T. Davis (A number of trees, thoroughly naturalized and apparently spreading).

Alnus incana (L.) Willd.   Grant City.

Quercus Brittoni Davis. Watchogue.

Salix fragilis latifolia And. Todt Hill.

Populus heterophylla L. Huguenot, W. T. Davis.

Tipularia unifolia (Muhl.) B. S. P. Tottenville.

Calopogon pulchellus R. Br. Watchogue, C. W. Leng.

Habenaria blephariglottis (Willd.) Torr. Watchogue, W. T. Davis.

Cypripedium acaule Ait. An albino form. Mrs. Heylyn.

Tradescantia Virginica L. Arlington, W. T. Davis.

Wolffia Columbiana Karst. Old Town Pond, Thos. Craig.

Potamogeton Spirillus Tuckm. Court House, Mrs. N. L. Britton.

Udora Canadensis Michx. Clove Lake, W. T. Davis.

Pinus echinata Mill. Linoleumville and not uncommon along the south side, W. T. Davis. Arrochar and Four Corners.

Eriophorum Virginicum L. Tottenville, W. T. Davis; Gifford's, Mr. Twiggs.

Panicum pubescens Lam. Richmond Valley.

Panicum microcarpon Muhl. Tottenville.

Panicum commutatum Schultes. Richmond Valley.

Panicum verrucosum Muhl. New Dorp, A. A. Tyler.

Carex tenera Dewey. New Dorp, A. A. Tyler.

Carex muricata L. New Dorp.

Carex Muhlenbergii Schk. Abundant at Richmond Valley.

Carex tribuloides Wahl. Tottenville and Grant City; A. A. Tyler.

Dryopteris cristata (L.) Gray. Oakwood, Mrs. N. L. Britton. Mariners' Harber.

Azolla Caroliniana Willd. Clove Valley, Thos. Craig. (Introduced some years ago by Mr. Samuel Henshaw.)

Salvinia natans (L.) All. Silver Lake and in a small pond on Ocean Terrace, Thos. Craig. (Probably introduced.)

ARTHUR HOLLICK,
N. L. BRITTON.